Quantum World

Quantum World

The Wave Nature of Our Universe

Dewey L. Boatmun

Library of Congress Control Number: 2010917584
ISBN: Hardcover   978-1-4568-2330-6
               Softcover   978-1-4568-2329-0
               Ebook   978-1-4568-2331-3

This book was printed in the United States of America.

To order additional copies of this book, contact:
Xlibris Corporation
1-888-795-4274
www.Xlibris.com
Orders@Xlibris.com
86553

Contents

To:
Danny,
Edward,
Julie,
& Wesley

Introduction

I am an old man—retired, but no longer disillusioned by it. My house is quiet, my kids grown, and I have had more than a little time to wonder about things. Imagine Thoreau and his experiment at Walden Pond. I am not that interesting, but I do live alone, in a house in the woods. I wrote the following pages in that solitude.

Generally, I wanted to put in writing those ideas which have occupied my mind over the last twenty years; specifically:

- Why is there something rather than nothing?
- What is "something"?
- What is "nothing"?
- What is the nature of universal expansion?
- Why is the mass density of the universe so close to the critical density?
- What is dark energy?
- What is quantum gravity?
- Does everything have a wavelike nature?
- Why are there no right-handed neutrinos?
- Why does time move preferentially in one direction?
- Why is there so much matter, and so little anti-matter?
- What is space?
- What is time?
- What is mass . . . .

My answers to these questions, in some instances, resulted in further questions, which I faithfully answer in this volume. In separate instances, the answers awe me still. I am amazed by the celestial mechanism; its simple complexity; its marvelous choreography, and the more deeply I probe its secrets, the more profoundly I'm affected by the answers. I'm sure that you know what I mean. We are the same, you and I.

I'm afraid, though, that because of my original intentions, my own mind may inadvertently block the reader's progress, for I had intended to share these ideas with the widest possible audience. This proved impractical over time. The minds of Susskind the Plumber and Joe the Plumber are not to be reconciled by Dewey the Tinkerer. I am not the man for that job. The initial chapters, then, you may pursue at your leisure; however, they were intended only to bring the laity up to speed. Please focus your criticism upon the later chapters concerning *cosmic logic, the principle of "nothing", the Planck wave, and quantum gravity*; it is for those ideas that I must be either complimented or condemned. I also ask indulgence for the fact that I am a mathematician, and while a capable physicist, I'm sure to prove an average historian. I'm even worse at story-telling. It's true. I didn't even write this introduction. After twenty years of tinkering, though, I'm through tinkering, and am resolved that I am either right, or I am crazy. Indulge an old man, and decide.

Dewey

Frank's Trucking

Frank's Trucking is a small company with a fleet of tanker trucks, which all have identical engines. Each engine delivers about 350 horsepower at 3000 rpm, and each truck is overhauled every 4,000 hours, so that, at the appropriate time, a truck is brought into the shop, the engine is removed and replaced with a rebuilt engine, and then the engine which was removed is subsequently rebuilt and installed in another truck. Likewise, alternators, starter motors, fuel tanks, pulleys, belts, and bearings are all removed, replaced, or rebuilt. Finally, the truck is refueled by the next truck to be overhauled, the refurbished truck is then hooked up to the incoming tanker trailer, along with any undelivered inventory therein, and is back on the highway; all very efficient, and economical. The complete job is called an overhaul package, and each package is meticulously itemized and tracked. As a result these overhaul packages are constantly being exchanged between trucks.

The company owner, a very professional and astute individual named Frank itemizes and categorizes company business in every way possible. For instance, since 4,000 hours at 350 horsepower and 3,000 rpm merits an engine change, he calculates horsepower hours per rpm

$$350X4000(hp)(hrs)/3000(rpm) = 467(hp)(hr)(min)$$

and keeps a computer log on every truck. Each package exchange involves two engines, which would mean

$$933(hp)(hr)(min)$$

per exchange and

$$467(hp)(hr)(min)$$

per truck, so that the number amount for every package exchange was a constant

$$k = 933(hp)(hr)(min)$$

and calculations involving truck overhauls or overhaul package exchanges would all involve some combinations of

$$nk$$

or

$$\frac{nk}{2}$$

Likewise, Frank keeps a close eye on fuel consumption which is itemized under the heading *Energy*, whereby he meticulously calculates energy consumed per mile driven. The constant (k) is an explicit statement of the total amount of energy consumed per period at a given rpm, or likewise an explicit statement of the total work done per period at a given rpm. In fact, using the constant (k) he could actually calculate the amount of work done per engine revolution.

One day, while meditating on how to make his computer log more informative, he decided to convert horsepower to kilowatts, so as to consolidate all his energy consumption. Also, out of curiosity he decided to convert hours and minutes to seconds. When he had finished, and was admiring his work, he had the strangest feeling that he had seen all this before, but couldn't quite put his finger on it. When expressed in these new dimensions, the constant (k) becomes

$$k = 7.52X10^{10}(joul)(sec)$$

Frank proudly labeled his equation "Frank's Constant".

At the same time, while analyzing fuel consumption per highway mile, Frank seemed to remember that the rate at which energy changes with respect to distance is called "force". In this case, he knew he was dealing with a force that propels trucks down the highway. He jokingly named it the "Frank Force", and noted that the total energy consumed (or equivalently the total work done) during any arbitrary period was equal to the Frank force times the total distance driven.

Well, the phone rang about that time. It was Frank's stock broker. Frank sold the business two weeks later, retired and moved to Florida.

Actually, by now you have probably realized that there is no Frank, no such trucking company, and no Frank's constant. But imagine if, for instance, an alien being was able to hack into Frank's computer. If he were able to decipher the contents therein, he would probably conclude that, on earth, elementary particles were called trucks, and that the forces between trucks were mediated by the exchange of what earthlings call packages. The exchanges are quantized as integer multiples of Frank's constant so that the total energy exchanged between trucks is equal to Frank's constant times the frequency of engine rotation.

Our universe also works by exchanging packages (little tiny packages). Horsepower hours are measured in joules, and rpm are measured in revolutions per second, so that the quantized value is also

$$(nk)(joule)(sec) \tag{1.1}$$

The constant (k), in this case is a very small number (6.63×10^{-34}). When factored as $6.63 \times 10^{-34}(joule)(sec)$ and denoted (h) it is called Planck's constant in honor of Max Planck, who in 1900 introduced the concept that the energy of an oscillation always comes in discrete packages, or *quanta*, and that the energy states of an oscillation are always an integer multiple of Planck's constant times the frequency of oscillation. The smallest possible energy change, therefore, is

$$\Delta E = h\nu \tag{1.2}$$

where ν (denoted by the Greek letter nu) is the frequency of oscillation.

In the real world the exchanged packages are called bosons and trucks are called fermions. Bosons and fermions are categorized according to spin state. Bosons have integer spin (nh), and fermions have half integer spin $(\frac{n}{2}h)$. Fermions include the quarks and leptons which form all the familiar elementary or composite particles of matter like electrons and protons.

Bosons include photons, the intermediate vector bosons, gluons, and the hypothetical graviton. Photons are the quanta of light which mediate the electromagnetic force. The intermediate vector bosons, W$^+$,

W⁻, and Z^0, mediate the weak force, the force responsible for some types of radioactive decay. Gluons mediate the strong force between quarks. And of course, the hypothetical graviton, if it exists, would mediate the gravitational force.

Light waves (or photons) are oscillations. This was well established by the time Max Planck made his conjecture about the discreet nature of oscillations. Interference experiments such as Thomas Young's famous double slit experiment in the early 1800s demonstrating constructive and destructive interference between light beams showed the wave nature of light. Later, in the most recent century, certain phenomena such as the photoelectric effect, X-ray absorption, and blackbody radiation made it appear that light sometimes behaved like a discrete particle rather than a wave. Even more confusing is the fact that elementary particles such as electrons sometimes behave like waves. For instance in some double slit experiments beams of electrons exhibit interference patterns, behaving like waves rather than discrete particles.

In 1924 a French physicist named Louis de Broglie proposed that matter particles have a dual nature, and that in addition to mass and velocity they also have an associated wavelength, and that, like electromagnetic waves, these should be related by Planck's constant:

$$\lambda = \frac{h}{mv} \tag{1.3}$$

where λ (the Greek letter lambda) is the wavelength and mv is the particle's momentum.

The first practical application of this hypothesis was to the energy states of the hydrogen atom (the simplest atom). In particular it clarified the spectra of electromagnetic emissions and absorption of hydrogen atoms in differing energy states.

Roughly speaking, an oscillation in one dimension is a straight line. Oscillations in two dimensions produce a circle or an ellipse, and oscillations in three dimensions result in a circle that wobbles or precesses about its central axis. Try to imagine one of those Halloween sparklers that you played with as a child. If you stand in the dark and move it back and forth from left to right, your friends would see something like a straight line. If you then bent your knees so as to move up and down at the same time in perfect synchrony, they might see something like an ellipse or a circle. And finally, if you were able to maybe twist back and forth while you were doing all this (without hurting yourself) they might see something like a wobbling circle. This is a gross oversimplification; however the orbital electron in the hydrogen atom does something like this, except unlike you and the sparkler the electron oscillations are quantized just as de Broglie hypothesized. With respect to each of the three oscillations the product of an electron's mass, velocity and orbital radius must always equal an integral multiple of Planck's constant i.e.

$$mvr = n\hbar \qquad (1.4)$$

or equivalently

$$I\omega = n\hbar \qquad (1.5)$$

where $I = mr^2$ and is called the moment of inertia and ω (denoted by the Greek letter omega) is the angular frequency (in radians per second). Here we are using $\hbar$ (pronounced h-bar) instead of h. $\hbar = h/2\pi$ and it is more convenient when measuring in radians rather than wavelength.

There are quantum numbers associated with each of the three oscillations. The quantum number n is associated with the oscillation in the first dimension, where n is proportional to the square root of the orbital radius. The quantum number ℓ signifies oscillations in two dimensions, where the electron moves with discrete amounts of angular momentum. The quantum number m_l refers to oscillations in three dimensions where, when placed in a magnetic field, the electron orbit wobbles, or precesses, about the orbital axis (this is called Larmor precession). These quantum numbers are integers and thus allow only discreet changes, i.e. $n\hbar$. And finally, the quantum number m_s refers to the actual spin of the electron itself and can only take on the values $(+1/2\hbar)$ and $(-1/2\hbar)$. The quantum changes in the electron orbits of atoms result in the electromagnetic radiation which we perceive as color in the world all around us.

It should be stressed that it is not just the motion of the electron that is wavelike, but that the actual nature of the electron is wavelike.

Now an elementary particle is more complicated than just a simple wave. Every known particle has a specific mass, specific charge, specific spin, etc. the origins of which are much more complex than the descriptions above, however all of these particle attributes are themselves wavelike in nature. In fact, we will see that the universe as a whole behaves in a wavelike manner.

The Special Theory of Relativity

By the late nineteenth century the wave nature of light was firmly established and it was widely assumed that, as with all other waves in nature, water waves, sound waves, waving strings, etc, there had to be some medium through which to wave. Some "thing" had to be waving, and it was widely assumed that this "thing" permeated all of space and it was given the name of luminiferous ether. This ether would therefore provide a universal frame of reference, a cosmic stage upon which Isaac Newton's laws of motion could be played out. All objects would move effortlessly through this ether without resistance or friction and vibrations of the ether would carry light over vast distances at remarkable speeds. Toward the end of the century the search was on to identify and measure in some way this remarkable luminiferous ether.

In 1897 two American researchers, Albert Michelson and Edward Morley, at what is now Case Western Reserve University in Cleveland, Ohio devised an ingenious and remarkably sensitive experiment. The experiment took advantage of the fact that the earth is moving with a relatively high velocity around the sun (about 100,000 km/hr). By measuring the velocity of light both parallel to and perpendicular to that motion they hoped to establish once and for all that elusive "universal frame of reference". The results were both shocking and perplexing.

The measured velocity seemed to always be the same no matter how it was measured. This result was bewildering. In an attempt to save the idea of an ether an Irish professor, George Francis FitzGerald, proposed that there was an interaction between a moving object and the ether, and that the length of the object was somehow shortened or contracted when moving in a direction parallel to the movement of the ether. This idea was also proposed independently and expanded upon by Hendrik Lorentz. Lorentz's mathematics led to a factor

$$\frac{1}{\sqrt{1 - \frac{v^2}{c^2}}} \tag{2.1}$$

now called the γ (gamma) factor which relates the velocity of the object and the velocity of light to the degree of contraction. This contraction is now called FitzGerald-Lorentz contraction. It wasn't until 1905 in a paper written by Albert Einstein, a young patent inspector working in the Swiss patent office, that what is now called special relativity was put in its present context. Einstein did away with the ether altogether, insisting that there was no preferred frame of reference; that all motion was relative, and that every observer could consider himself to be at rest. This required that distance, time, and mass were all variables and that the speed of light was a constant for all observers regardless of their state of motion.

To explain time dilation Einstein imagined two observers in different frames of reference, one (O') in a boxcar on a train moving down the tracks, and the other (O) standing at the station. At the floor of the boxcar is a light source that flashes a pulse of light toward a mirror mounted on the top of the car, at a distance H from the light source. Now (O) and (O') will observe two different situations (See Fig. 1), and remember that both observers must find the same value for the speed of light. (O') must see the pulse of light moving in a straight line from the source to the mirror and then back to the source, a distance 2H so that

$$t'_2 - t'_1 = \frac{2H}{c} \tag{2.2}$$

(O), on the other hand, sees the light traveling at an angle along the hypotenuse of the first triangle and then back down along the hypotenuse of the second triangle, a distance $(2c\frac{(t_2-t_1)}{2})$.

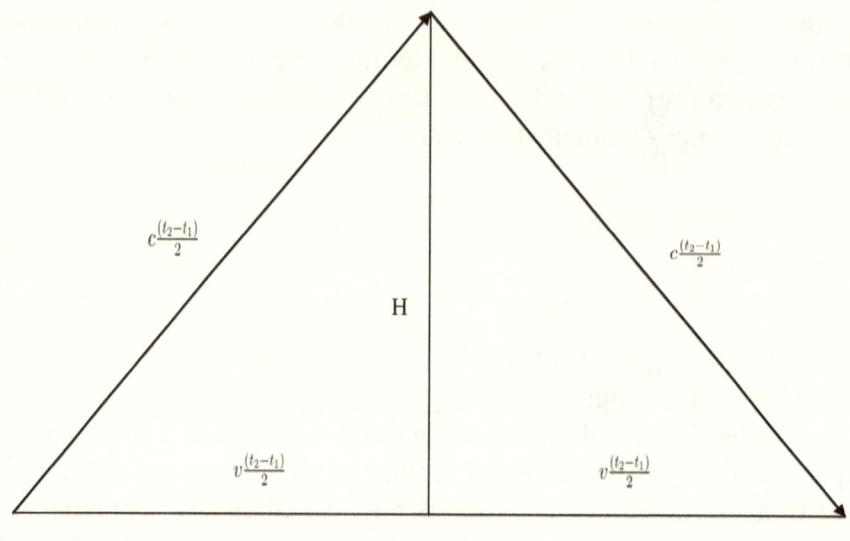

FIG. (1)

The Pythagorean Theorem says that

$$[\frac{c}{2}(t_2 - t_1)]^2 = H^2 + [\frac{v}{2}(t_2 - t_1)]^2 \qquad (2.3)$$

$$H^2 = [\frac{c}{2}(t'_2 - t'_1)]^2$$

so that

$$[\frac{c}{2}(t'_2 - t'_1)]^2 = [\frac{c}{2}(t_2 - t_1)]^2 - [\frac{v}{2}(t_2 - t_1)]^2 \qquad (2.4)$$

and

$$(t_2 - t_1) = \frac{(t'_2 - t'_1)}{\sqrt{1 - \frac{v^2}{c^2}}} \qquad (2.5)$$

so that in order for (O) and (O') to find the same value for the speed of light regardless of their state of motion, each must measure time differently.

    If (O')'s clock is running more slowly than (O)'s, it presents another problem. Suppose that (O') takes off his shoe and drops it from distance H. He must see it hit the floor with a momentum $p_{shoe} = m_{shoe}v_{shoe} = m_{shoe}\frac{H}{t}$. (H) is the same for both observers however each measured a different (t). The law of conservation of momentum is every bit as dearly held as the constancy of the speed of light. For both to hold then each must find a different value for m_{shoe}. In retrospect, this is really not all that surprising, in view of Einstein's most recognized equation, namely the equivalence of energy and mass.

$$E = mc^2$$

which means that energy and mass are equivalent, since c (the speed of light) is a constant. In other words, in order to propel the shoe down the track in the first place (not to mention the train), mass had to be converted into energy (coal to steam for instance) and the portion of energy required to propel the shoe must manifest itself as the shoe's additional mass.

If all this is true, then why aren't we constantly resetting our watches? Why aren't airline pilots constantly arguing with ground controllers about the distance from Washington to Seattle, etc.? It's just that even velocities which we consider extremely high, such as those of missiles or fighter aircraft, are infinitesimally small compared to the speed of light. Consider for example an earth satellite. The escape velocity at the earth's surface, the minimal velocity necessary to achieve orbit is about 25,000 miles per hour.

$$\frac{25,000mi/hr}{186,000mi/sec} = \frac{25,000mi/hr}{669,600,000mi/hr} = 0.0000373$$

which is less than 1 one hundredth of 1 percent of the speed of light.

Suppose the following. NASA launches a manned mars mission. The distance to Mars is (l_o meters). The mass of the space craft (not including fuel) is (m_o kilograms). In addition, after reaching maximum velocity the crew synchronizes its clock with mission control's clock. The flight time to Mars, as measured by mission control is t_0. Without going into too much detail this requires that the distance to Mars, as measured by the flight crew, is less than l_o meters, its mass as measured by mission control is greater than m_o kilograms, and the flight crew's clock is running more slowly than that of mission control, albeit ever so slightly. In particular, at any point thereafter:

$$l = l_o\sqrt{1 - \frac{v^2}{c^2}} \qquad (2.6)$$

$$m = \frac{m_0}{\sqrt{1 - \frac{v^2}{c^2}}} \qquad (2.7)$$

and

$$t = t_o\sqrt{1 - \frac{v^2}{c^2}} \qquad (2.8)$$

Although counterintuitive, there is extensive experimental evidence confirming the laws of special relativity. Time dilation, for instance, was tested (*Hafele and Keating, Nature 227 (1970), pg 270 (proposal). Science Vol. 177 pg 166-170 (1972) (experiment)*) using atomic clocks flown aboard commercial airliners and synchronized with an earthbound clock at the US Naval Observatory, and was found to agree with the predictions of special relativity.

If (Eq. (2.6), which describes FitzGerald-Lorentz contraction, is put in the form

$$l^2 = l_0^2\left(1 - \frac{v^2}{c^2}\right)$$

then

$$\frac{l^2}{l_o^2} + \frac{v^2}{c^2} = 1 \qquad (2.9)$$

In addition, if c is measured in light seconds so that $c = 1$, and l_0 is measured in units (1 foot, one yard, one astronomical unit etc.) Eq. (2.9) becomes

$$\frac{l^2}{1} + \frac{v^2}{1} = 1$$

or

$$l^2 + v^2 = 1 \qquad (2.10)$$

which defines the unit circle shown in Fig. (2). A unit circle has continuous rotational symmetry; i.e. no matter how you rotate it, it looks the same. For every continuous symmetry in nature, there is a conserved quantity. For example, the invariance of a system of equations under translation of spatial coordinates is responsible for conservation of momentum. The invariance of the energy of a system under translation in time insures conservation of energy, and symmetry under rotation through a fixed angle is responsible for conservation of angular momentum. The conserved quantities in Fig. (2), represented by the hypotenuse of the triangle, are the length of the measuring stick at Mission Control (l_0) and the speed of light (c). The height of the triangle represents the space craft velocity (v), and the base of the triangle represents the length of the measuring stick in the space craft (l) (from mission control's point of view). As the velocity approaches the speed of light (when the arrow will point directly to the 12 o'clock position), the length of the space craft's measuring stick, and the space craft itself, approach zero. Likewise the astronaut's clock slows to a halt and the mass (m) of the craft approaches infinity. The craft can never actually reach the speed of light, since that would require an infinite amount of energy, but notice (and keep in mind) that on the other side of the light barrier l becomes ($-l$)

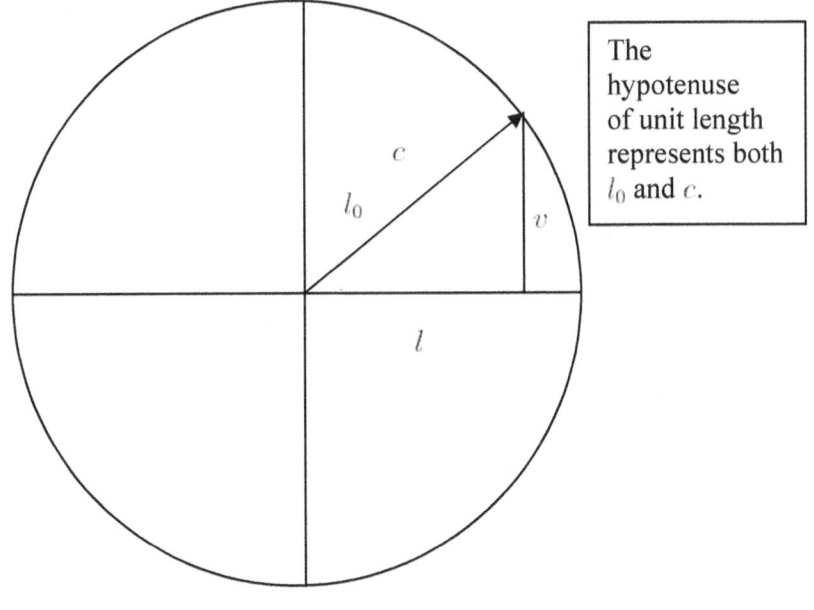

The hypotenuse of unit length represents both l_0 and c.

Fig. (2)

The General Theory of Relativity

The term special relativity is "special" in the sense that it applies only to uniform motion. In 1916 Einstein published a paper which "generalized" relativity to include non-uniform, i.e. accelerated motion.

An important element of Einstein's theory of general relativity is the principle of equivalence. In particular, that inertial forces and gravitational forces are indistinguishable. Einstein imagined a person in a rocket ship in outer space accelerating at a rate of 9.8 meters per second per second, which of course, is the acceleration due to gravitation at the earth's surface. The person inside would be held to the floor of the ship just as if it were still on the launching pad. In other words gravitational acceleration and inertial acceleration are one and the same. The predictions of general relativity maintained that space and time are distorted by a gravitational mass, and in fact, this distortion is what we actually perceive as gravitation.

The first test of general relativity was carried out in 1919 by British astronomers Sir Frank Dyson and Sir Arthur Eddington. General relativity predicted that the path of light from a star would be bent when it passed near a massive object. The most massive object in our solar system is, of course, the sun. The only time the sun can be observed is (naturally) in the daytime, so Dyson and Eddington undertook expeditions to observe the sun during a total eclipse and to compare the positions of stars near the disc of the eclipsed sun with the known positions of these stars. The stars, though visible, appeared slightly out of position, showing that light does, in fact, travel in a curved path near a massive object.

General relativity also predicts that time slows down in a gravitational field. A clock at sea level runs more slowly than a clock on Mount Everest, albeit ever so slightly Other experiments carried out later in the century, such as observations of binary pulsars, confirmed the predictions of general relativity to a greater degree of accuracy.

Remarkably, proof of space-time curvature actually preceded Einstein's 1916 paper. The precession of the perihelion of the planet Mercury had been a longstanding problem in celestial mechanics. Most planetary orbits are elliptical to some extent, with the perihelion of the orbit being the planet's closest approach to the sun. The precession of the perihelion of the orbit of Mercury, the planet nearest to the sun, disagreed with calculations by about 43 seconds of arc per century and had defied all attempts to account for it. As it turns out Mercury is so close to the sun's mass that the effects of general relativity on space and time are actually observable, and the relativistic calculations agreed with observation exactly.

More recently, other predictions of general relativity, such as gravitational red-shift, radar echo delay from massive objects, and gravitational lensing have also been observed. In fact, anyone who has a global positioning system in his or her auto is constantly conducting experiments in special and general relativity, and is confirming the two theories each time a route is successfully navigated. It happens that the system requires such accuracy that time dilation due to special and general relativity must be taken into account. For instance, the U.S. global positioning satellites orbit at a distance of 26,560 kilometers above the center of the earth. Meanwhile, the surface of the earth (sea level) is about 6,360 kilometers above the center of the earth. General relativity maintains that an atomic clock on a GPS satellite at 26,560 kilometers will run faster than one on the surface of the earth. Specifically:

$$t = t_0 \sqrt{1 - \frac{2Gm}{rc^2}} \qquad (3.1)$$

where G is the gravitational constant, c is the speed of light, r is the clock's distance from the center of the earth, and t_0 is time as measured at an infinite distance from the earth's mass. As a result, an atomic clock on the satellite will run faster than one on the earth's surface. The difference is minute however, only about $45 \mu sec/day$.

In addition, while the satellite clock will run faster than the earthbound one due to the effects of general relativity, that same clock will run more slowly than its earthbound cousin due to the effects of special relativity. In particular

$$t = t_0 \sqrt{1 - \frac{v^2}{c^2}}$$

where v is the satellite's orbital velocity (about 3.9 kilometers per second). This effect is smaller, though: only about $7 \mu sec/day$. The net difference then, is about

$$(45 - 7) = 38 \mu sec/day$$

Therefore the satellite and earthbound atomic clocks must be able to reconcile a $38 \mu sec/day$ difference.

Although difficult to test, because of the extreme weakness of the gravitational force on the small scale, the general theory of relativity has been highly successful, and is generally accepted in the scientific community.

Matter Waves

A hypothetical particle constructed of waves, as proposed by de Broglie, is referred to as a "wave group" or "wave packet". When discussing waves it is sometimes convenient to speak in terms of "frequency" (denoted ω—the Greek letter omega—and measured in radians per second), and "wave numbers" (denoted k, where $k = \frac{2\pi}{\lambda}$). When two waves (A and B), moving in one dimension at the speed of light but differing in frequency are superimposed, they form a wave group or packet, as in Fig. (3), which differs slightly from either A or B. In particular, the envelope (A+B) moves at a subluminal velocity and this is called the group velocity (v) of the wave group. The wave enclosed within the envelope moves at a superluminal velocity and that is called the phase velocity (u).

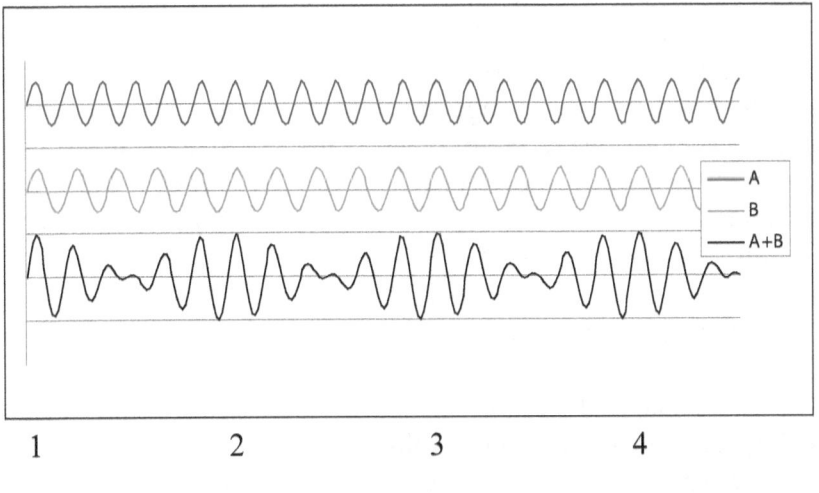

FIG. (3)

The group velocity and phase velocity are therefore, respectively (inversely) and (directly) proportional to the speed of light. Since the phase velocity $u = r\omega = \frac{E}{h}\frac{h}{mv}$ we have

$$u = \frac{mc^2}{h}\frac{h}{mv} = \frac{E}{p} = \frac{c^2}{v}$$

so that

$$uv = c^2$$

With respect to frequency (ω) and wave number (k),

$$v = \frac{d\omega}{dk}$$

and

$$u = r\omega = \frac{\omega}{k}$$

The group velocity is the actual velocity of the particle being represented by the wave group.

Let me just say, at this point, that the only knowledge of calculus necessary to understand the ideas expressed herein is the following:

The term above ($\frac{d\omega}{dk}$) is called the derivative of ω with respect to k. The derivative of any variable, for example y (a function of x), with respect to x is simply the change in y with respect the change in x, as the change in x approaches zero. In other words a derivative measures rate of change. Meanwhile, the inverse of a derivative is called an integral (and denoted $\int dy$ or $\int dx$), where the integral of dy is simply the sum of all infinitesimal changes in y between two limits, and integral of dx is simply the sum of all infinitesimal changes in x between two limits. Therefore, if dy is an infinitesimal part of y then $\int dy = y$, and of course, $\int dx = x$.

In this context therefore, if waves of different frequencies are superimposed, then the group velocity $v = \frac{d\omega}{dk}$ is the difference in frequency with respect to the difference in wave number and the phase velocity, $u = \frac{\int d\omega}{\int dk} = \frac{\omega}{k}$.

Now a wave, such as A or B in Fig. (3), has a well defined momentum determined, as we said before, by Planck's constant. In particular, the momentum

$$p = \frac{\hbar}{r}$$

where

$$r = \frac{\lambda}{2\pi}$$

A look at A or B, though, tells you nothing about the particle's position. There is an infinite uncertainty about its position. A+B however says that the particle may be located at position 1, or position 2, or position 3, or position 4 . . . . or position n, which is still a large uncertainty, but only about half as uncertain as it was before. However, there is now uncertainty about the momentum. The momentum of (A+B) is either that of A or of B. If we add more and more waves to the superposition the envelopes representing the particle will get narrower and narrower and taller and taller; and the distance between them will get greater and greater. Therefore, the uncertainty in the particle's position will have gotten smaller, but at the expense of a greater uncertainty in the particle's momentum. If an infinite number of waves were added, only one very tall and narrow envelope would exist in the entire universe, and we would have absolute certainty in its position, yet an infinite uncertainty in its momentum.

This is known as the Heisenberg uncertainty principle, named for German physicist Werner Heisenberg. The principle states

$$\Delta p \Delta x \approx \hbar \qquad (4.1)$$

where Δp is the uncertainty in the particle's momentum and Δx is the uncertainty in the particle's position along the x axis. The same principle assigns uncertainty in a particle's energy state, specifically,

$$\Delta E \Delta t \approx \hbar \qquad (4.2)$$

where ΔE is uncertainty in energy, and Δt is uncertainty in time.

Just as waves can exhibit particle-like behavior, particles can exhibit wavelike behavior. Refer back to the gamma factor mentioned in the chapter on special relativity.

$$\frac{1}{\sqrt{1 - \frac{v^2}{c^2}}}$$

As was mentioned above,

$$u = \frac{mc^2}{h}\frac{h}{mv} = \frac{E}{p} = \frac{c^2}{v}$$

and

$$uv = c^2$$

Making the substitution

$$\frac{1}{\sqrt{1 - \frac{v^2}{c^2}}} = \frac{1}{\sqrt{1 - \frac{v^2}{uv}}} \tag{4.3}$$

$$= \frac{1}{\sqrt{1 - \frac{v}{u}}} \tag{4.4}$$

and recalling that $v = \frac{d\omega}{dk}$ and $u = \frac{\int d\omega}{\int dk}$

$$\gamma = \frac{1}{\sqrt{1 - \frac{\frac{d\omega}{dk}}{\frac{\omega}{k}}}} = \frac{1}{\sqrt{1 - \frac{\frac{d\omega}{dk}}{\frac{\int d\omega}{\int dk}}}} \tag{4.5}$$

If we consider $\frac{\int d\omega}{\int dk}$ in Eq. (4.5) to be the set of all individual $\frac{d\omega}{dk}$'s, another interesting phenomenon of relative motion will have been clarified. Michael Faraday (1791-1867) was the first to establish a unification of two seemingly different forces of nature: electricity and magnetism. In particular, this unification implies that electricity and magnetism are simply two different manifestations of the same force, and as with the theory of special relativity, the principle of relative motion is central to the argument.

Again, just as waves are superimposed to form particles, particles as we all know, can be superimposed to form waves. The most familiar of course, are water molecules forming waves upon the surface of a pool. The simplest way to make electrons move in waves is to force them through a coil of copper wire by applying an electrical potential across the ends of the wire. Better yet, wrap the copper coil around an iron bar and apply the electrical potential. This turns the bar into a permanent magnet with electron orbits in the iron atoms aligning themselves with the electrons moving through the coil. This alignment will remain even after removing the copper coil. Electric charges moving in such waves produce what is called a magnetic induction (B), the magnitude of which is determined by the *Biot-Savart Law*,

$$B = \frac{\mu i dl sin\theta}{4\pi r^2} \tag{4.6}$$

where $\mu = 4\pi X 10^{-7} weber/amp \bullet m$. In the situation where we are talking only about a coil of copper wire, i is the electrical current through the wire, and dl is the length of the wire. In the situation where we're considering a permanent magnet, i is the electrical current consisting of all the trillions upon trillions of aligned electrons orbiting in the iron atoms, and dl is the sum total of all the electrons orbital radii. The sum of all electron velocities ($\frac{\int d\omega}{\int dk}$) = u can of course exceed c (the speed of light).

Actually since $\mu = 4\pi X 10^{-7} weber/amp \bullet m$ Eq. (4.6) reduces to

$$B = \frac{10^{-7} i dl sin\theta}{r^2} \tag{4.7}$$

Meanwhile the electric field intensity for a charge Q is determined by Coulomb's Law and is given by

$$E = \frac{1}{4\pi \epsilon} \frac{Q}{r^2} \tag{4.8}$$

$\frac{1}{4\pi\epsilon} = c^2 X 10^{-7}$ so that Eq. (4.8) becomes $\frac{c^2 X 10^{-7} Q}{r^2}$.

Remembering that $c^2 = uv$ and treating u and v as vectors, we take uv to be the cross product of u and v. A simple example of a cross product is torque, as in the torque that results from your foot applying force to a bicycle peddle, which would be denoted $\vec{F} X \vec{r}$. In that instance $\vec{F} X \vec{r} = |\vec{F}| |\vec{r}| \sin\theta$, where the brackets signify absolute value, F is force, r is the length of the peddle bar, and θ is the angle your leg makes with the bar.

For the vectors u and v, $\vec{u} X \vec{v} = |\vec{u}| |\vec{v}| \sin\theta = |\frac{\int d\omega}{\int dk}| |\frac{d\omega}{dk}| \sin\theta$. With this in mind Eq. (4.8) becomes

$$E = \frac{10^{-7} u Q v \sin\theta}{r^2} = (10^{-7})\left(\frac{\int d\omega}{\int dk}\right) Q \left(\frac{d\omega}{dk}\right) \frac{\sin\theta}{r^2} = \frac{(10^{-7})(i)(dl)(\sin\theta)(v)}{r^2} \quad (4.9)$$

where $(\int d\omega)(Q) = i$, and $\frac{1}{\int dk} = dl$.

But

$$\frac{(10^{-7})(i)(dl)(\sin\theta)(v)}{r^2} = \frac{(4\pi X 10^{-7})(i)(dl)(\sin\theta)(v)}{4\pi r^2}$$

$$= \frac{\mu i dl \sin\theta}{4\pi r^2}\left(v\right)$$

which is Eq. (4.6), the *Biot-Savart Law*, multiplied by velocity (v). This says that an electric field (E), is equivalent to a magnetic field (B), moving at a velocity (v). This indicates that electricity and magnetism are simply two different manifestations of the same phenomenon. What has happened is that rotational symmetry was broken when we forced electrons through the copper wire coil, in either a left-handed or right-handed manner. In doing so we defined a preferred direction in space as evidenced by the sudden appearance of the angle (θ) in Eq. (4.9).

From Eqs. (4.7) and (4.9)

$$E = B\left(\frac{d\omega}{dk}\right) \tag{4.10}$$

so that

$$\frac{E}{B} = \frac{d\omega}{dk} \tag{4.11}$$

Eq. (4.11) says that a change in an electric field with respect to a change in a magnetic field is nature's way of keeping track of relative motion.

Think of it this way: in the case of a permanent magnet, the iron bar consists of a huge number of iron atoms some of whose orbital electrons have been aligned so as to move with circular wavelike motions. Just as each individual bird in a flock flies at some particular velocity, each individual orbital electron is moving at some particular orbital velocity, and the sum of all these orbital velocities is $\frac{\int d\omega}{\int dk} = u$. Meanwhile, just as the flock itself is flying at some particular "flock" velocity, the entire group of orbital electrons (i.e., the bar itself) can be moved at some particular "group" velocity, $\frac{d\omega}{dk}$.

The velocity at which the iron bar moves ($\frac{d\omega}{dk}$), becomes the rate at which a magnetic field (B) changes with respect to an electric field (E). For instance, if the magnetic field (B) of the magnet is moved across a segment of copper wire, and the velocity of the group of electron waves in the magnet (i.e., group velocity) is $\frac{d\omega}{dk}$, it induces an electric field (E) in the wire, and therefore the rate at which the induced electric field changes with respect to the changing magnetic field is $\frac{d\omega}{dk}$. This, by the way, is known as an electric generator.

Inversely, we could have applied an electric potential across the segment of copper wire. This would in turn cause the magnet to move (rotate actually), changing the angle θ. This is called an electric motor.

In these two configurations, $\frac{\int d\omega}{\int dk} = u$ is a constant (the sum of all electron orbital velocities). If we, on the other hand, replace the magnet with a coil of copper wire we can artificially rotate the wire coil by sinusoidally varying the current through the coil, thus changing the angle θ. In this case, the relative motion ($\frac{d\omega}{dk}$) remains constant (as long as the AC potential across the coil remains constant), and both E and B vary sinusoidally. This, by the way, would be called an electrical transformer;

like one of the big barrel looking devices that you sometimes see on utility poles.

An obvious question at this point might be, "what mediates this force that exists between, for instance, a wire segment and a permanent magnet in an electric motor"? Well, just as the exchange of packages between Franks trucks mediated a mysterious force between those trucks, an exchange of massless packages (photons, or light waves, in this case) mediates a force between the wire segment and the permanent magnet. Moving through space at $c = 299,792,458$ meters per second, this field of photons constitutes an electromagnetic field.

While one of Frank's truck engines constituted a rotation in real space, a photon constitutes a rotation in complex space; part real and part imaginary. Just as a changing electric field E in the electric motor described above produces a changing magnetic field B, a photon constantly exchanges energy between its electric and magnetic fields, and the rate at which E changes with respect to B is exactly 299,792,458 meters per second. Since $uv = c^2$, it follows that $u = v = c$ for the photon. If $\frac{d\omega}{dk}$ for the photon is equal to the set of all $\frac{d\omega}{dk}$'s, then the set must consist of just one $\frac{d\omega}{dk}$, one discreet package of electromagnetism.

The 2nd Law of Thermodynamics

The laws of thermodynamics came to be formulated as a direct result of the industrial revolution. In particular, engineers continually endeavoring to improve the efficiencies of steam engines proposed certain principles which evolved over time and eventually came to be considered "laws"—"laws" of nature.

The first systematic formulation of the principles appeared in Sadi Carnot's *Reflections on the Motive Power of Fire* in 1824, and by the end of the century the theory had been developed in great detail by researchers such as Rudolf Clausius, William Thompson, and Ludwig Boltzmann, among others.

The first law has to do with conservation of energy and simply states that "energy can neither be created nor destroyed, it can only change forms".

The second law, the most intriguing one, states that "the entropy of an isolated system, not at equilibrium, will tend to increase over time, approaching a maximum value at equilibrium".

The third law states that "the entropy of a system approaches a constant minimum as the temperature approaches absolute zero".

The third law in fact establishes the principle of absolute zero. The second law will be the one focused on herein.

The "entropy" of an isolated system can be thought of roughly as the degree of disorder of the system. Isolated systems in nature tend to go from lower entropy toward higher entropy as time passes, that is to say as time goes forward. Examples abound, but the few narrated below will give the general flavor of the argument.

1. On Saturday morning, a child's mother orders him to clean his room. He looks at the room with a sense of foreboding. There

The 2nd Law of Thermodynamics

are dirty clothes lying about, socks, shirts, etc. There are books scattered here and there, maybe some candy wrappers, crayons, and the like. The bed is unmade. After some diligent effort, everything is put in its proper place, books are put back in the book case, the bed is made, and the room looks great. It is in a well ordered state. Next Saturday morning, however, the situation has reverted. In the span of a week the room has changed from a well ordered state back into a highly disordered state, from a system of low entropy to a system of higher entropy. The low entropy state can be recovered but it will take work, and work requires energy. Energy is required to decrease the entropy of a system.

2. Take an air conditioner or a refrigerator as an example. A typical air conditioner or refrigerator consists of the following: (1). a compressor motor, (2). a compressor coil, (3). an expansion valve, (4). a cooling coil, and (5). a reservoir to hold the Freon or other coolant. When the device is activated, the compressor motor pumps Freon into the compressor coil. The increased pressure causes the Freon (and subsequently the coil) to heat up. This heat is dissipated into the cooler surrounding air. The compressed and cooled Freon is then vented through the expansion valve and into the cooling coil where, because of decreased pressure due to expansion, the Freon (and subsequently the cooling coil) cool further. We now have a highly ordered system. All the cold Freon has been moved into the cooling coil and all the hot Freon has been moved into the compressor coil. All the slow moving molecules went to the cooling coil and all the fast moving molecules to the compressor coil. It's fascinating when you think about it. It required energy though to achieve this low entropy state. If the device is turned off and the compressor motor stops, the hot and cold Freon will, as time moves forward, converge toward equilibrium, with the hot Freon getting cooler and the cold Freon getting warmer, an increase in entropy.

3. Imagine a large room, maybe an auditorium; yes let's make it an auditorium, filled with people. Everything is fine until a skunk enters through the door at the rear of the auditorium. As you can imagine, panic breaks out: screaming, shouting, climbing over seats. This terrifies the skunk and it activates its self-defense mechanism as the crowd cowers at the front of the auditorium. I

won't go so far as to say that this is a well ordered system but we do have the following facts. The foul smelling air is concentrated near the rear of the auditorium, while all the breathable air is concentrated near the front. Of course that's all about to change and all because of the entropy mandates of the second law of thermodynamics. That's right; as time moves forward, the skunk air and breathable air are required to mix together toward a point of equilibrium and a state of maximum entropy, not to mention disorder.

The most frustrating thing about these, and all other examples of the second law, is that the underlying Newtonian laws of motion are all time reversal invariant (TRI). In other words, the laws of motion are exactly the same for any action and its time reversed equivalent. For instance, imagine two basketballs colliding in midair. It would be impossible to tell a collision in forward time from a collision in reverse time. The mechanics of the situation insist that time, on the average, should stand still, and yet it is overwhelmingly apparent that it does not. Time marches on, as they say, but only in one direction.

Consider the following, though. Instead of a skunk, let's just say that an aromatic liquid is placed in a cup in the middle of a room, and the aroma quickly fills the entire room. Shortly afterward, imagine gradually slowing, then stopping, and then finally reversing the trajectory of every atmospheric molecule in the room which would be tantamount to a time reversal. In this scenario the past should be every bit as entropic as the future, with the aroma, once reversed, re-concentrating into its original position inside the cup, diffusing back throughout the room in reverse time, and eventually reaching equilibrium and refilling the entire room. So why isn't it just as easy for time to move backward rather than forward? It could be argued, and rightly so, that if each molecule were reversed (a time reversal), then before every molecule re-concentrated into the cup, one would see the door open and a person walk backwards into the room, pick up the aromatic liquid container, walk backward out of the room, and close the door, thereby rescuing the second law. This is frustrating.

It has been argued that cosmology lies at the heart of the entropy paradox. This view maintains that the universe began in a state of extremely low entropy and that entropy must therefore increase with time. Of course, if entropy increases with time then it must have been lower in the past. Perhaps it was exactly zero at time zero. It is perhaps a reasonable assumption, but it's not completely clear what zero entropy entails. For instance, does it imply some sort of crystalline type structure? Zero entropy is sometimes represented, for purposes of argument, as a perfect crystal at a temperature of absolute zero (-273° Celsius). If the universe began in the fiery explosion called the big bang, the perfect crystal at absolute zero scenario seems a little far fetched. At the same time, the irreversible increase in entropy with the passage of time seems indisputable, so that a less entropic past is also undeniable.

The consideration of the subject of entropy from a statistical standpoint was introduced in 1870 with the work of the Austrian physicist Ludwig Boltzmann. Statistically speaking, isolated systems tend in the direction with the most available microstates. Specifically, the entropy S of a macrostate is defined as follows:

$$S = k_B ln \Omega$$

where

S is the entropy of the macrosystem
k_B is the Boltzmann constant
ln is the natural logarithm
and
Ω is the number of available microstates in the specific macrostate.

From this perspective, the constant tendency toward greater entropy (and consequently the constant direction of the arrow of time) is simply a matter of probability. Accordingly, it is not impossible for waves in a pool of water to suddenly converge to a point and eject a diver, or glass shards to come together in the shape of a window and eject a baseball, or billiard balls to come together into a triangle and eject a cue ball; but it is highly improbable.

In fact, if the flow of time depends on one's state of motion, as was discussed in the section on special relativity, then perhaps relativity and the 2nd law are related. If we look back at Eq. (4.5)

$$\gamma = \frac{1}{\sqrt{1 - \frac{d\omega}{dk}\frac{\omega}{k}}} = \frac{1}{\sqrt{1 - \frac{\frac{d\omega}{dk}}{\int \frac{d\omega}{dk}}}}$$

and rewrite it as

$$t = t_0 \sqrt{\frac{\frac{\int d\omega}{\int dk} - \frac{d\omega}{dk}}{\frac{\int d\omega}{\int dk}}}$$

and consider $\frac{\int d\omega}{\int dk}$ to be the set of all $\frac{d\omega}{dk}$'s, then, as we know time speeds up in the moving frame of reference as $\frac{\int d\omega}{\int dk}$ gets larger and $\frac{d\omega}{dk}$ gets smaller. Does the set of all $\frac{d\omega}{dk}$'s represent the number of available microstates? After all, adding energy (larger $\frac{d\omega}{dk}$) would result in a reduction in the number of microstates ($\frac{\int d\omega}{\int dk}$), and energy, after all, is required to reduce the asymmetry of forward time over reverse time. Is the 2nd law of thermodynamics determined by the laws of relativity?

While the forward progress of time and the constant increase in entropy are certainly obvious, surely our perception of these phenomena is one of averages. Imagine, for example, a breezy day. The wind is blowing at 10 mph out of the southwest. If you possessed a tiny device that was capable of measuring individual air molecule velocities and recording those velocities, it might start with a print-out such as

1. 273° @ 943 mph
2. 057° @ 315 mph
3. 162° @ 621 mph
4. 045° @ 14 mph
5. 114° @ 1327 mph
6. 233° @933 mph

(with true north being 0°)

Averaged over time, however, the wind would be blowing out of the southwest @ 10 mph. Considering the discussion which follows, it would seem that the speed and direction of the passage of time is something like this.

In 1993 a theorem by Evans et al called the "fluctuation theorem" predicted that for small systems and short time-periods, entropy was actually consumed as well as produced. The theorem also quantified that consumption and production. In particular

$$\frac{P(\sum_t = A)}{P(\sum_t = -A)} = e^{At} \tag{5.1}$$

which simply says that the *asymmetry* in time increases exponentially with the *passage* of time. Some years ago an experiment at The Australian National University, conducted by Wang, Sevick, Mittag, Searles, and Evans actually confirmed the theorem. For small systems and short time-periods, time actually does move in both directions.

    The second law is truly one of the most mysterious and yet compelling truths of nature. The fact that we can remember the past but not the future, that we can change the future but not the past, and that the arrow of time has something to do with steam engines and disorder is to say the least, bewildering. I think I'll go sit in front of the air conditioner and think about it.

Time, Charge, and Parity

Time

The first asymmetry of nature to become apparent was, in fact, that of time, blatantly and sometimes painfully so. Even primitive animals must posses at least a rudimentary sense of time. For example, a predator pursuing prey must be capable of anticipating future circumstances, future motions, future positions, etc. Migrating animals have evolved to anticipate changing climates. Squirrels store up nuts for the winter.

Most creatures learn from the past and use knowledge of the past to improve their future. Humans are certainly aware of the passage of time. Most human endeavor could probably be characterized as using knowledge of the past to project the future. In fact, the most dramatic advances in the history of Homo Sapiens have seemingly been sparked by improved methods of recording the past. Human speech itself, for example, story telling etc.; passing on to future generations the lessons learned by past generations. Then came the written word, chiseled in stone so to speak, a much more accurate method of recording the past. Pen and paper are a big improvement over stone and clay. Next the printing press, vastly increasing the number of stories that can be told and passed on into the future. Now, most recently we enter the computer revolution, vastly increasing the speed and volume of information retrieved from the past, and improving the speed and accuracy of future projections.

Even our measurement of time betrays our preoccupation with the asymmetry of time and its irreversible march in one direction. No one knows exactly when, but at some point, our ancestors perceived the relationship between frequency and time and settled on the human heartbeat as the smallest increment of time. The inverse square of the heartbeat frequency became our hour. The inverse of the orbital frequency of the moon became our month. The inverse of the orbital frequency of the earth's orbit around the sun became our year. They were obviously aware of the relationships among the year, the growing season, the harvest season, and the dormant season for the sole purpose of manipulating the future and insuring their survival. Even microbial colonies seem to learn from the past, evolving over time with changing environments (drug resistant bacteria, for example).

However just as the passage of time is so obviously asymmetric, the Newtonian laws of motion are time reversal invariant (TRI). These laws work just as well at predicting the past, given initial conditions, as they do at predicting the future, given the same conditions; however, they invariably predict the future, and record the past.

Charge

Until 1928, it apparently never occurred to anyone that the presence of so much matter implied an asymmetry of nature. Matter just "was". Its existence was undeniable, and while there were obviously questions of "why" and "how much", apparently no one suspected that each particle of matter had an additive inverse. In that year the existence of antiparticles was implied by the negative energy solutions of P. A. M. Dirac's relativistic wave equation. A particle of antimatter would have the same amount of mass as its matter counterpart, but all quantum numbers would be equal and opposite. This asymmetry of matter over antimatter is called a "charge violation" since particles of matter and antimatter have equal and opposite electrical charges. Antimatter was first observed in 1932 when C. D. Anderson observed positrons (the antiparticle of the electron) in cosmic rays from outer space. Today the existence of antimatter is an accepted fact and it is produced routinely in particle accelerator experiments. Antimatter particles don't last long, though. They mutually annihilate when they come in contact with their additive inverse, their matter counterpart.

Parity

There is an intuitive relationship between time and space. A particle moving from left to right is the time reversed equivalent of a particle moving from right to left. Likewise, a wave moving from left to right is the time reversed equivalent of a wave moving from right to left. Just as the Newtonian laws did not seem to discriminate between forward time and reverse time, it was believed that these laws did not discriminate between right and left. That all changed in the mid-1950s. Tsung Dao Lee and Chen Ning Yang searching through scientific literature could find no direct evidence for conservation of parity in interactions involving the weak force. In experiments in 1956-57 it was found that, when cobalt-60 atoms were aligned by the presence of a magnetic field, beta decay, via the weak force, exhibited a special preference. In particular, decay occurred preferentially in one direction over the other.

The three spatial coordinates are usually represented graphically as in Fig. (5)

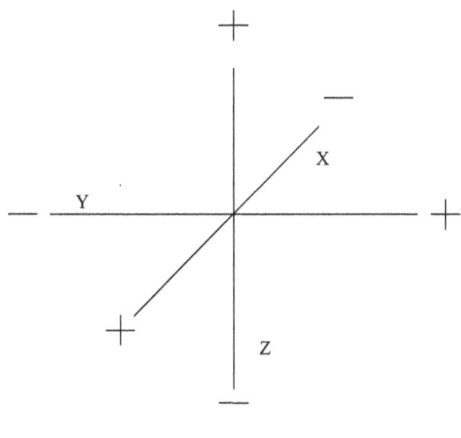

FIG 5

Three-dimensional space can be represented vectorially as right-handed space and left-handed space, that is by

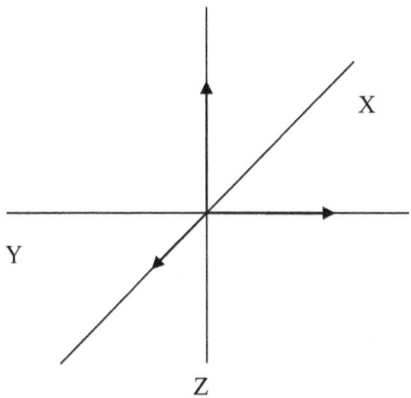

and

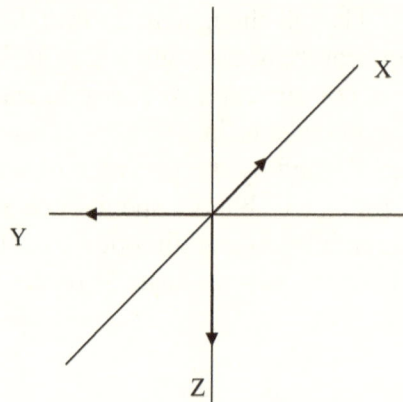

respectively. One is the mirror image of the other. In the top image, if you point the right-hand in the (+x) direction, and bend the fingers in the (+y) direction, the thumb will point in the (+z) direction. The same operation performed on the bottom image with the left-hand in the (-x) direction, and the fingers bent in the (-y) direction will point the thumb in the (-z) direction.

Once it was shown that parity conservation was violated in weak interactions, it was assumed that at least the combined symmetry of charge and parity (CP) was conserved. In experiments in 1964, however, Christenson, Cronin, Fitch, and Turlay, while observing the decay of neutral K mesons, discovered that these mesons could transform into their antiparticles and vice versa but not with the same probability. There also seemed to be long lived and short lived versions. While observing the decay of long lived neutral K mesons, the following decay mode was observed.

$$K_L \longrightarrow \pi^+ + \pi^-$$

If CP were a good symmetry, the K_L would have CP=1 and could only decay to three pions, not two, therefore the above decay mode violates the combined symmetry of charge and parity.

CPT

It is now generally agreed that the overall symmetry of charge, parity, and time (CPT) is a good symmetry. In fact the Lüders-Pauli CPT theorem expresses exactly that. A time violation is, in fact, equivalent to a charge parity violation, and must be accompanied by a reverse charge parity violation if (CPT) is to be conserved. For instance, if charge symmetry were a perfect symmetry, then there would be equal amounts of matter and antimatter in the universe and the two would mutually annihilate. Neither matter nor antimatter would exist. If parity were a perfect symmetry, then right-handed space and left-handed space would cancel. Space would not exist. If time were a perfect symmetry, then time would move backward at the same rate that it moves forward. Time would stand still.

If any of the three symmetries is broken however, then in order to preserve perfect CPT symmetry, there must be a combined violation of the other two symmetries in order to compensate. If a fluctuation in the early universe resulted in a space filled with matter (a CP violation) there would need to be a compensating time violation, with time moving preferentially in one direction.

Conversely, it would seem that the reduction of a combined symmetry violation should result in a reduction of the third violation. For instance, an infinite amount of mass would need to be converted into energy in order for, say an astronaut, to be accelerated to the speed of light. As this was happening the astronaut would have to see forward moving time slowing down, left-handed space shrinking, and the amount of mass (presumably in the form of matter) in the rest of the universe decreasing. Interestingly enough, if we refer back to Fig. (2), we see that, on the other side of the light barrier, space, time, and mass become negative. This begs the question, "if there were a sudden time reversal (a reversal of the (T) violation), would all matter be suddenly transformed into anti-matter, with all neutrinos spinning right-handedly"?

In fact, the CPT theorem maintains that if there existed a mirror image of our own universe, which happened to be filled with anti-matter, and where time moved backward, that universe would evolve using the same physical laws as our own. Newton's laws would apply, the standard model of particle physics would apply, and special and general relativity would apply.

Asymmetrics

The notions of space, time, and mass as were redefined by the laws of special and general relativity, might at first, seem counterintuitive. In fact, it must be remembered that what we perceive as space, time, and matter, are, in each case, asymmetries; the asymmetry of forward time over reverse time; the asymmetry of matter over anti-matter; and the asymmetry of left-handed space over right-handed space. Recall the fluctuation theorem (Eq. (5.1))? Remember the experiment that confirmed that theorem, in which it was shown, that for small systems and short periods of time, time actually does flow backward?

A familiar example of an asymmetry, which most of us are familiar with, would be a bank balance. One can either have a positive balance or a negative balance, that is to say, either an asymmetry of income over outflow, or vice versa; positive dollars over negative dollars or vice versa. A balance of zero would represent perfect symmetry. (Well, maybe "perfect" is a poor choice of words). Anyway, one might spend one's money on "fast cars and fine clothes", in which case, "fast cars and fine clothes" would be equivalent to negative dollars; and inversely, a large positive bank balance would probably be indicative of slower cars and cheaper clothes. Just as a charge and parity violation would constitute a time violation, "fast cars and fine clothes" would constitute a dollar violation. In fact, in this case, "dollars" are how "fast cars and fine clothes" are measured. The two are inversely proportional, just as relativistic mass is inversely proportional to relativistic distance and relativistic time, and from a relativistic point of view, mass is how space and time are measured (and vice versa). When a relativistic body decelerates, it gives up mass in exchange for space and time. When it accelerates, it gives up space and time in exchange for mass. When the exchange rate approaches infinity, the inverse exchange rate approaches zero.

Inverses are very important at a fundamental level because all measurements, scientific or otherwise, are relative. To illustrate multiplicative inverses, for example, imagine a stick that is 4 feet long. If you break the stick at a point three fourths of the distance from one end, and then use the short stick to measure the long stick, and the long stick to measure the short stick, the measurements will be

$$3 \text{ feet/yard}$$

and

$$1/3 \text{ yard/foot}$$

respectively. The product of the two measurements

(3 feet/yard) x (1/3 yard/foot)

is equal to the identity element for multiplication, i.e. 1.

Just as mass can be exchanged for space and time, space can be exchanged for time. In fact space is officially measured relative to time. One meter is defined as the distance light would travel in a vacuum in exactly $\frac{1}{299,792,458}$ of a second. The speed of light, on the other hand, is officially 2,99,792,458 meters/second.

($\frac{1}{299,792,458}$ seconds/meter) x (299,792,458 meters/second)=1, the identity element.

In subsequent chapters we will see that, in a very natural way, mass is, in fact, defined in terms of space and time.

Inverses are so common in our everyday life that we often fail to recognize them. The price of gasoline on a given day might be 3 dollars/gallon. This is just another way of saying that the value of a dollar is 1/3 gallon of gas.

(1/3 gallon/dollar) x (3 dollars/gallon)=1

And as the value of a gallon of gas approaches infinity, the relative value of the dollar approaches zero.

Examples abound: the price of hamburger, oil, the number of cents/dollar, dollars/euro, the number of feet/mile, etc. etc. Some inverses, such as the price of gasoline, for example, are variable, while others, cents per dollar, for example, are constants. Examples such as these invariably involve some sort of bidirectional exchange where one thing of value moves in one direction and the other thing of value moves in the opposite direction, usually across some sort of barrier, real or imagined, a grocery counter for example.

More on these bidirectional exchanges later.

The Standard Model

The unification of three of the four recognized forces of nature is one of the greatest accomplishments in physics and the culmination of decades of effort. The three include the strong, weak, and electromagnetic forces. The fourth recognized force (gravitation) is not included in the unified model.

Gravity and electromagnetism, though, are the most obvious of the four, and are responsible for almost all of the observable phenomena of everyday life. The gravitational force confronts us constantly, both as friend and foe. We struggle against it constantly, yet life as we know it would be impossible without gravity.

Electromagnetic phenomena can be much more subtle than those of gravitation however. Being responsible for light itself, they do result in everything we actually "see", so that in this respect electromagnetism is strikingly obvious. A little less obvious might be a phenomenon such as a baseball hitting a bat. In this instance the force between the bat and ball is actually an electrostatic force between the orbital electrons in each of the two objects, which come very close, but do not actually touch, at what we perceive as impact. Electromagnetic forces are also responsible for chemical properties so that we owe to them our very bodies, our senses, our thoughts.

More subtle still are the strong and the weak forces, both of which manifest themselves at very small distance scales, in atomic nuclei for example. The strong force holds atomic nuclei together while the weak force is responsible for some types of radioactive decay. As it turns out, the electromagnetic and weak forces are actually two different manifestations of a more fundamental force, now called the electroweak force. The unification of the strong, weak, and electromagnetic forces, and the elementary particles between which they act, is part of what is called the *standard model* of particle physics.

At the heart of the standard model is the principle of symmetry. We are all familiar with various examples of symmetry. Reflective symmetry, the symmetry which our body possesses, which makes us look like our mirror image. Translational symmetries, repetitive patterns along a line for instance, are common in things like wallpaper. Repetitive notes in a musical piece exhibit translational symmetry. Rotational symmetries are common in everything from spinning wheels to hockey pucks.

There are discreet symmetries, a repeating pattern on wallpaper, for instance, and there are continuous symmetries, a rotating circle, for example. As was pointed out earlier, "for every continuous symmetry in nature, there is a conserved quantity". For example, the invariance of a system of equations under translation of spatial coordinates is responsible for conservation of momentum. The invariance of the energy of a system under translation in time insures conservation of energy, and symmetry under rotation through a fixed angle is responsible for conservation of angular momentum.

Symmetry is so important in mathematics that there is a special branch of mathematics devoted to it, called group theory. In particular, for any set G which is a subset of a set S, G is a group with respect to some specific operation ($*$) (such as multiplication, addition, etc.) if there is an equality relation and

1. G is closed under $*$ (that is, if x and y are in G then $x * y$ is in G)
2. $*$ is associative (that is, for all x, y, z in G, $x * (y * z) = (x * y) * z$
3. G has an identity element, e, such that $x * e = e * x = x$ for all of G
4. G contains inverses. (that is for each element a of G, there exists an element b of G such that $a * b = b * a = e$

For example, the set of all positive real numbers is a group with respect to multiplication since

1. If x and y are positive real numbers then x times y is a positive real number.
2. If x, y, z are positive real numbers, then $x * (y * z) = (x * y) * z$
3. G has an identity element 1, since $x * 1 = 1 * x = x$ for all of G

4. For any positive real number a there exists a positive real number $b = \frac{1}{a}$, such that $a * b = a * \frac{1}{a} = 1$

However this set of positive real numbers is not a group with respect to addition since it has no additive identity and no additive inverses.

Different groups have different representations, and oftentimes unrelated and seemingly disparate operations will have identical group representations. One may not necessarily know the exact nature of the operation but will know the exact rules it will follow. Different operations which have identical group representations are said to be "isomorphic".

For instance suppose three students, (1) Joe, (2) Bob, and (3) Bill, decide to play a trick on a new teacher. The teacher has a seating chart but doesn't know any of the students. Below is shown all the possible seating arrangements with e (the identity element) being the proper seating. For instance, the τ permutation would be (3) Bill pretending to be (1) Joe, (1) Joe pretending to be (2) Bob, and (2) Bob pretending to be (3) Bill.

$$
\begin{array}{lll}
e: \begin{array}{l} e(1) = (1)\text{Joe} \\ e(2) = (2)\text{Bob} \\ e(3) = (3)\text{Bill} \end{array} &
\tau: \begin{array}{l} \tau(1) = (3)\text{Bill} \\ \tau(2) = (1)\text{Joe} \\ \tau(3) = (2)\text{Bob} \end{array} &
\gamma: \begin{array}{l} \gamma(1) = (3)\text{Bill} \\ \gamma(2) = (2)\text{Bob} \\ \gamma(3) = (1)\text{Joe} \end{array}
\end{array}
$$

$$
\begin{array}{lll}
\rho: \begin{array}{l} \rho(1) = (2)\text{Bob} \\ \rho(2) = (3)\text{Bill} \\ \rho(3) = (1)\text{Joe} \end{array} &
\sigma: \begin{array}{l} \sigma(1) = (2)\text{Bob} \\ \sigma(2) = (1)\text{Joe} \\ \sigma(3) = (3)\text{Bill} \end{array} &
\delta: \begin{array}{l} \delta(1) = (1)\text{Joe} \\ \delta(2) = (3)\text{Bill} \\ \delta(3) = (2)\text{Bob} \end{array}
\end{array}
$$

Now, however, suppose the new teacher is suspicious and decides to call roll. The three students can still continue the ruse and confuse the teacher even further by answering with a different name.

The set forms a group with respect to the operation ○ (where the operation ○ can be construed to mean "followed by"), and we can derive what is called a multiplication table for the group by doing composition mapping. For instance, the composition mapping for the situation where Joe sits in Bob's seat and answers "here" when Bill's name is called (Joe pretending to be Bob pretending to be Bill) would be

$$
\rho \circ \sigma = \begin{array}{l} \rho(\sigma(1)) = \rho(2(1)) = (3) = \text{Bill} \\ \rho(\sigma(2)) = \rho(1(2)) = (2) = \text{Bob} = \gamma \\ \rho(\sigma(3)) = \rho(3(3)) = (1) = \text{Joe} \end{array}
$$

as indicated in the multiplication table below. In other words, if you locate the symbol ρ in the left column, and then its intersection with the symbol σ in the top row, the point of intersection is labeled γ.

Incidentally, it turns out that τ is identical to $\rho \circ \rho$, so that τ is usually signified ρ^2.

$\circ$	e	ρ	ρ^2	σ	γ	δ
e	e	ρ	ρ^2	σ	γ	δ
ρ	ρ	ρ^2	e	γ	δ	σ
ρ^2	ρ^2	e	ρ	δ	σ	γ
σ	σ	δ	γ	e	ρ^2	ρ
γ	γ	σ	δ	ρ	e	ρ^2
δ	δ	γ	σ	ρ^2	ρ	e

This group is known as the symmetric group of order 3. It might not seem at first that this mischievous scheme has much in common with an equilateral triangle. As it turns out, though, the set of all possible rotations of an equilateral triangle about a given axis which can be made without changing the image of the triangle is also a group, in fact, the symmetric group of order 3, and can also be represented by the same multiplication table. These two operations, therefore, are said to be isomorphic.

The gauge group $SU(3)XSU(2)XU(1)_Y$ of the standard model serves to unify the strong, weak, and electromagnetic interactions. Here SU, and U refer to types of matrices. U stands for unitary, S stands for special, and the numbers (1, 2, and 3) refer to the number of real dimensions. $U(1)_Y$ is the matrix associated with the complex unit circle; in other words, for a unit circle in two "real" dimensions, the x and y coordinates of any point on the circle correspond to $(1 \cos \theta)$ and $(1 \sin \theta)$, respectively, for any arbitrary angle θ. In complex space with one "real" and one "imaginary" dimension the corresponding coordinates would be $(1 \cos \theta)$ and $(i \sin \theta)$, where $i = \sqrt{-1}$. Just as the symmetry of a rotation in real space is called spin, there is an internal symmetry in complex space called isospin, and the complex space in which it lives is called isospace.

A unitary matrix is a complex matrix whose conjugate transpose is also its inverse. $U(1)_{EM}$ describes the electromagnetic interaction. $U(1)_{EM}$ has 1 boson associated with it—the photon—the exchange of which mediates the electromagnetic force. Likewise $U(1)_{EM}$ has 1 conserved quantity, or quantum number, and that is electrical charge.

    A special unitary matrix is special in the sense that its determinant is exactly equal to 1. For special unitary matrices (SU(n)), there are $(n^2 - 1)$ matrices and each matrix corresponds to a particular gauge boson (force exchange particle). In addition there are $(n - 1)$ diagonal matrices, each of which corresponds to a conserved quantity, or quantum number.

For SU(2) then we have

$(2^2 - 1) = 3$ matrices known as Pauli matrices,

$$\sigma_1 = \begin{bmatrix} 0 & 1 \\ 1 & 0 \end{bmatrix}$$

$$\sigma_2 = \begin{bmatrix} 0 & -i \\ i & 0 \end{bmatrix}$$

$$\sigma_3 = \begin{bmatrix} 1 & 0 \\ 0 & -1 \end{bmatrix}$$

corresponding to 3 bosons ....W^+, W^-, Z^0

and $(n - 1) = 1$ diagonal matrix
corresponding to....................................z component of isospin

For SU(3) we have

$(n^2 - 1) = 8$ matrices.
corresponding to 8 bozons ....8 gluons
(These three dimensional matrices,
called Gell-Mann matrices, are
simply the Pauli matrices with
extra rows and columns of zeros.)

and $(n - 1) = 2$ diagonal matrices

corresponding to...2 color coordinates carried by
each gluon

The symmetry $SU(2)XU(1)_Y$, however, is spontaneously broken in much the same way that the rotational symmetry of our iron bar (in chapter 4) was broken when it became magnetized, and the iron atoms aligned. It acquires a preferred direction in space. With this broken symmetry, 4 massless gauge bosons, A^1, A^2, A^3 and B of the symmetric theory are converted into the 3 massive gauge bosons of the weak interaction and the massless photon, A, of the electromagnetic interaction.

Two of the massive bosons, the W^+ and W^- are two different linear combinations of A^1 and A^2, and are electrically charged. The third massive boson and the massless photon are electrically neutral and come about as a result of a mixing of A^3 and B at an angle of about $30°$ called the Weinberg angle, or weak mixing angle (θ_W). The masses of the W and Z (M_W and M_Z, respectively) are determined by this weak mixing angle. Specifically:

$$\cos \theta_w = \frac{M_W}{M_Z} \qquad (7.1)$$

Likewise, the coupling constants, g for $SU(2)$ and g' for $U(1)_Y$

$$\frac{g'}{g} = \tan \theta_w \qquad (7.2)$$

Elementary particles between which the forces interact are called fermions, and the carriers of these forces are called bosons. In much the same way that Joe, Bob, and Bill changed identity by exchanging seats and names, fermions can change identity by exchanging bosons. Fermions spin with half integer angular momentum in units of Planck's constant. Bosons possess integer spins. Fermions are divided into quarks and leptons. There are three generations of each, as shown below, with each successive generation being more massive than the previous.

	1	2	3		
	FERMIONS				BOSONS
QUARKS	u up	c charm	t top		γ photon
	d down	s strange	b bottom		g gluon
LEPTONS	ν_e electron neutrino	ν_μ muon neutrino	ν_t tau neutrino		Z neutral
	e electron	μ muon	t tau		W charged

The quark "flavors" of charm (C), strange (S), bottom (B), and top (T), while colorfully named, are actually conserved values resulting from group symmetries. All of these quarks have significantly different masses, and are therefore considered singlets; i.e. different particles. The up and down quarks have very similar masses and are considered to be "isotopic doublets", much like different isotopes of certain elements are distinguished by different atomic weights. The combined values of these conserved properties along with another called baryon number (A) yields yet another quark property called hypercharge (Y), such that

$$Y = A + S + C + B + T$$

Furthermore, the value of hypercharge (Y) along with the value of the z component of isospin (T_3) (not to be confused with topness) determine the electrical charge of any bound state of quarks. In particular, for the electrical charge (Q)

$$Q = \frac{1}{2}(A + S + C + B + T) + T_3$$

Every Quark possesses another charge fancifully called color. There are three colors (red, green, blue) and three anti-colors (anti-red, anti-green, anti-blue). Individual quarks do not exist in isolation, but form bound states with other quarks through the exchange of gluons. These bound states are called hadrons and hadrons themselves are further subdivided into mesons and baryons. Mesons are bound states of a quark and an anti-quark. Baryons are bound states of three quarks. Hadrons can only exist in combinations which are "colorless", i.e. white. For example a meson might have one red quark and one anti-red quark. A baryon might have one red, one green, and one blue (the combination of which would, of course, be white). Quarks do not carry integer electrical charge. The up, charmed, and top quark carry a charge of $+\frac{2}{3}$ and the down, strange, and bottom quarks carry a $-\frac{1}{3}$ charge. The familiar proton, for example, consists of two up quarks and one down quark for a charge of $2(+\frac{2}{3}) + (-\frac{1}{3}) = +1$. A neutron consists of two down quarks and one up quark for a charge of $2(-\frac{1}{3}) + (+\frac{2}{3}) = 0$.

Just as there are two quarks in each generation, there are likewise two leptons. The first generation of leptons contains the familiar electron and the electron neutrino. The second generation leptons are the muon and the muon neutrino, while the third generation consists of the tau and tau neutrino. The electron, muon, and tau carry a charge of -1. The three neutrinos are electrically neutral (hence the name neutrino). In addition, neutrinos have extremely small masses (if any mass at all).

The massless photon, the light wave, is the carrier of the electromagnetic force and is exchanged between electrically charged particles. The massless gluon is the carrier of the strong force and is exchanged between quarks. Gluons each carry two color indices (think seat location and name). The W^+, W^-, and Z^0 are massive bosons and mediate the weak force.

By exchanging gluons, a quark can transform into a quark of a different color. By exchanging W bosons, quarks and leptons can transform into quarks and leptons of different masses and different electrical charges, such as an up quark transforming into a down quark or vice versa, or an electron transforming into a neutrino or vice versa. Finally, exchanging photons or Z bosons goes full circle and changes a particle into itself. The exchange of bosons between fermions is responsible for the three forces of the standard model.

In summation, the gauge group $SU(3)XSU(2)XU(1)_Y$ of the standard model and the fermions and bosons defined by the group representations account quite nicely for three of the four known forces. General relativity, on the other hand, has supplied a very convincing description of gravitation. Quantum theory and the standard model, however, offer descriptions on the microscopic level, while general relativity describes gravitation on the macroscopic level. Thus far, a quantum theory of gravity has proven frustratingly elusive.

For instance, a relationship between two coupling constants (dimensionless constants which determine the strengths of the forces) can be expressed geometrically as with Eq. (7.2) above, however there is a huge problem when trying to show a geometric relationship between gravity and the forces of the Standard Model. Take gravitation and electromagnetism for example. The strength of the electrostatic force between the electron and proton in a hydrogen atom is on the order of 10^{39} times greater than the gravitational force between the two. In fact, if I were to try, here on this page, to represent the strength of these two forces as the two orthogonal legs of a right triangle, as in Eq. (7.2), the leg representing the electrostatic force would not fit inside the known universe. Even if the vector representing gravity were on the order of the diameter of an atom, the vector representing the electrostatic force would still not fit. In fact, the length of the vector representing gravity would have to be about 1836 times smaller than the hydrogen atom; about the distance by which the proton's position is perturbed by the orbiting electron.

This, it has been speculated, was not always the case, however. At distances on the order of 10^{-35} meters, it seems that the two forces are of comparable strength. In fact, some physicists believe that at that distance all the forces of nature may converge into a single force. This distance is referred to as the Planck distance (again in honor of Max Planck) and is the radial length of a very high frequency wave. The energy/mass of the wave is called the Planck mass, and the inverse of the Planck frequency is called the Planck time.

What seems to make this particular frequency special is that the gravitational force between two Planck waves seems to agree with quantum theory-at this frequency and no other. In fact, at times before the Planck time, present theories of physics and cosmology seem to break down.

Cosmic Inflation

By the second half of the twentieth century the concept of a big bang universe was on pretty sound footing. The discovery in 1929, by Edwin Hubble, that the universe was uniformly expanding in every direction led quite naturally to the assumption that the time reversed equivalent was a universe that uniformly contracted from every direction toward a singularity of sorts.

In addition there was mounting evidence for the big bang scenario from different corners. First of all, calculations carried out by Fred Hoyle (who, by the way, coined the term "big bang" in a moment of jest while debunking the theory) and others confirmed that this scenario would very accurately account for the already well documented relative abundances of hydrogen, helium, and lithium isotopes.

Secondly, if big bang theory was correct, a cosmic microwave background radiation was predicted by George Gamow, Ralph Alpher, and Robert Herman. This radiation would have been released in the distant past when the universe went through a phase transition and became transparent to radiation. In 1964, that background radiation was discovered, quite by accident by Arno Penzias and Robert Wilson at Bell Laboratories. While doing communications research they were plagued by an omnipresent source of background noise which seemed to come equally from every direction, and which they were unable to eliminate. As it turned out, this noise was in fact the predicted background radiation which had cooled as the universe expanded over the last 13.7 billion years to a temperature of about 2.7° Kelvin.

There were, however, still daunting difficulties with the big bang theory. For one, why did the universe appear so flat? For another, why was it so lumpy on the small scale yet homogeneous on the large scale? Why was it isotropic on the large scale, or the same in every direction?

Finally, most versions of big bang theory predicted the existence of huge numbers of magnetic monopoles: however none have ever been observed.

In 1980, a theory was proposed by Alan Guth which seemed to obviate many of the problems with the big bang. The theory was called "cosmic inflation", and while there are difficulties with the theory, many physicists feel that some version of cosmic inflation will eventually be confirmed.

At the heart of inflation theory is: (1) the idea of negative energy, and (2) the equivalence of mass and energy. If you shoot an arrow into the air, it requires energy. When that arrow reaches its maximum height, it possesses a certain amount of gravitational potential energy. Gravitational potential energy is considered negative energy. When the arrow starts to fall back to earth, some of that potential energy is converted into the kinetic energy of the arrow. Kinetic energy is considered positive energy. As the kinetic energy becomes more and more positive, the gravitational potential energy becomes more and more negative so that the total energy never changes. In other words, total energy is conserved in accordance with the first law of thermodynamics.

Since mass and energy are equivalent, as required by the theory of relativity, it was suggested first by Peter Bergmann, a physicist and associate of Albert Einstein, that mass in the universe represents positive energy and that gravitational potential energy, being negative, may exactly cancel out all the positive mass energy in the universe, positive energy being the additive inverse of negative energy.

Cosmic inflation postulates that, very early in the universe, fluctuations produced a period of very high energy density, but negative pressure. This negative pressure, in essence, would have resulted in a negative gravitational field. Negative gravity theoretically had the practical effect of causing an exponential growth in the diameter of the universe. This exponential growth would seem to solve the flatness problem in much the same way that the surface of a balloon gets flatter as it inflates. It might also help to resolve magnetic monopole problems. Magnetic monopoles would be few and far between.

Since gravity is equivalent to negative energy/mass, negative gravity would have to be equivalent to positive energy/mass, so that as the radius of the universe grew exponentially, theoretical particles could have acquired mass through what is called the Higgs mechanism. In a scenario which has been likened to a black hole turned inside out, instead of devouring matter from the surrounding universe, it would be belching out massive matter particles into an expanding universe.

The Higgs mechanism, so named for Peter W. Higgs of the University of Edinburgh, is a theoretical mechanism by which previously massless particles (known as Goldstone Bosons) acquire mass. For a massless particle moving at the speed of light, one real dimension (say the x dimension for instance) does not exist, since $l_x = l_0\sqrt{1 - \frac{c^2}{c^2}}$. The spin with respect to another massless particle cannot change, since three dimensions are required to define spin. Chiral symmetry exists. On the other hand, a particle with mass can change its relative motion with respect to another massive particle, whereby its spin can become either left-handed or right-handed. Chiral symmetry is broken. At the same time, a massless particle (such as a photon or light wave) represents what is called a transverse wave, meaning its oscillations are orthogonal (or at right angles) to its direction of propagation. It has no oscillation in the direction of its propagation. A sound wave, on the other hand, is a longitudinal wave. Air molecules passing a point along its direction of propagation are alternately denser and less dense. It has a longitudinal component of oscillation.

At very low temperatures, some materials become superconductors, conducting electricity without resistance. In superconductors, electric and magnetic field oscillations become short ranged, adding an extra degree of freedom, a longitudinal component. The Higgs mechanism postulates a super cooled condensate of charged massless particles, which acquire that extra degree of freedom through interaction with the non zero Higgs field

A massless particle which acquires a longitudinal component would exhibit an effective mass. Consider, for example, photons trapped between two mirrors (somewhat like a laser or a maser). In this case the photon field has acquired a longitudinal component. If one were to accelerate the mirrors parallel to the line of photon propagation, photons striking the leading mirror would be red shifted and photons striking the trailing mirror would be blue shifted in the same way that the radio waves from a trooper's radar gun are either red shifted or blue shifted, and how he or she knows to give you a ticket. (If this is not so, then try explaining that to the trooper.) The momentum of the blue shifted photons would be greater than the momentum of the red shifted photons resulting in an opposition to acceleration (the defining characteristic of mass).

The Higgs mechanism postulates a superconducting field of charged massless particles in the early universe which had become super cooled due to universal expansion, and that the vacuum expectation value (VEV) of the field is non zero. According to the most accepted version, four massless scalar bosons interact with this non zero Higgs field, whereby they acquire specific masses, depending on their angle of alignment. The first three become the W^-, W^+, and Z^0 intermediate vector bosons of the electroweak force, and the fourth becomes the yet undiscovered Higgs boson. The angle of alignment is known as the Weinberg angle θ_W , for the physicist Stephen Weinberg. Specifically $\cos\theta_W = \frac{M_W}{M_Z}$, where M_W is the mass of the W boson, and M_Z is the mass of the Z boson.

In the same way, quarks and leptons would acquire mass through interaction with the Higgs condensate.

Cosmic Logic

In statistical analysis and probability theory, diagrams used to demonstrate logical functions are called Venn diagrams, as in FIG. 6.

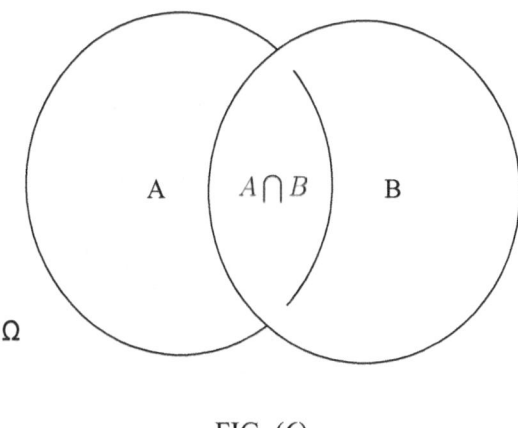

FIG. (6)

For two events A and B (elements of a sample space Ω), the probability of A is represented by the area of the left-hand portion of the figure, the probability of B is represented by the area of right-hand portion of the figure, and the probability of A and B (referred to as the "intersection" of A and B, and denoted $(_A \bigcap _B)$, or sometimes (AB), is represented by the overlapping portion. The probability of A or B (referred to as the "union" of A and B, and denoted by $(_A \bigcup _B)$ is equal to the probability of A plus the probability of B minus the probability of A and B, i.e.

$$P(_A \bigcup _B) = P(A) + P(B) - P(_A \bigcap _B) \qquad (9.1)$$

Of course, if A and B do not overlap, then A and B are said to be independent and $A \bigcup B$ is simply $(A + B)$.

Computer circuits perform logical functions using what are called logic gates. These gates operate as follows, where a 1 represents a truth and a 0 represents a falsehood.

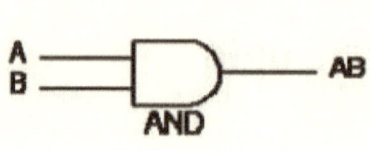

2 Input AND gate		
A	B	A.B
0	0	0
0	1	0
1	0	0
1	1	1

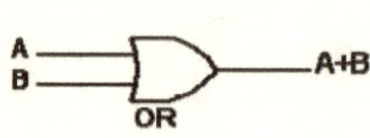

2 Input OR gate		
A	B	A+B
0	0	0
0	1	1
1	0	1
1	1	1

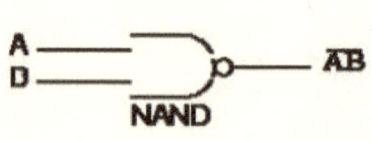

2 Input NAND gate		
A	B	$\overline{A.B}$
0	0	1
0	1	1
1	0	1
1	1	0

Here the bar over the AB signifies the negation of (A and B) ((i.e. not (A and B)).

Just as electrons can be made to perform logical operations, so can waves. Let wave $\vec{A} = \overrightarrow{\cos \theta_A} + \overrightarrow{\sin \theta_A}$, wave $\vec{B} = \overrightarrow{\cos \theta_B} + \overrightarrow{\sin \theta_B}$, and wave $\vec{C} = \overrightarrow{\cos \theta_C} + \overrightarrow{\sin \theta_C}$ be three unit vectors corresponding to rotations in the x-y plane. Let their relative phase angles be

$\theta_A = 150°$
$\theta_B = 30°$
$\theta_C = 270°$

as in Fig. (7)

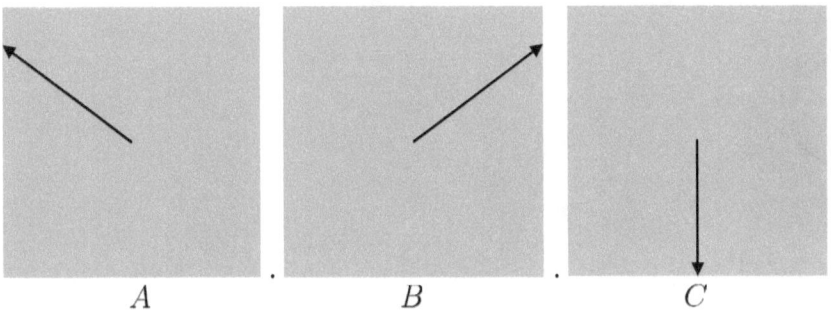

A B C

FIG. (7)

The vector sum of the two
unit vectors A and B is a
unit vector (AB, i.e. ($_A \bigcap _B$))
at 90°.

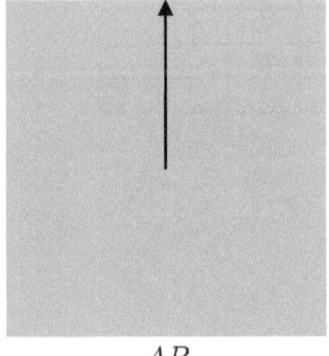

AB

FIG. (8)

Likewise the vector sum
of the two unit vectors
A and C is the unit vector
(AC, i.e. ($_A \bigcap _C$)) at 210 °.

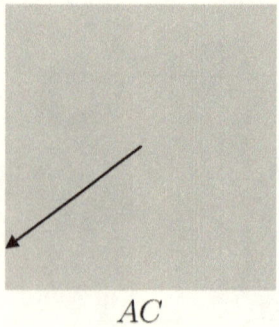

AC

FIG. (9)

It can be seen that the vector C at 270° would represent the negation of the unit vector (AB, i.e. $(_A \cap _B)$) at 90°, in other words:

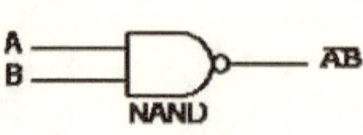

2 Input NAND gate		
A	B	$\overline{A.B}$
0	0	1
0	1	1
1	0	1
1	1	0

so that

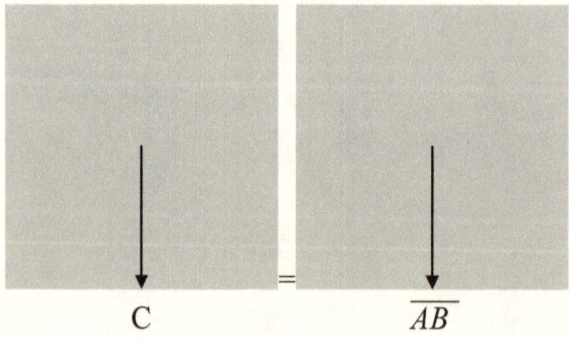

C $\overline{AB}$

FIG. (10)

And of course a unit vector at 30° would represent the negation of (AC, i.e. $(_A \cap _C)$).

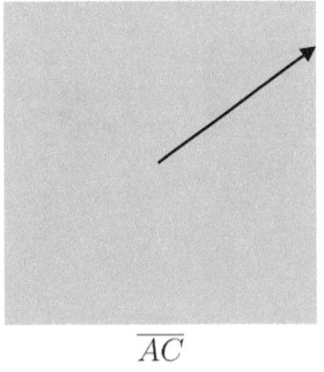

$\overline{AC}$

FIG. (11)

so that

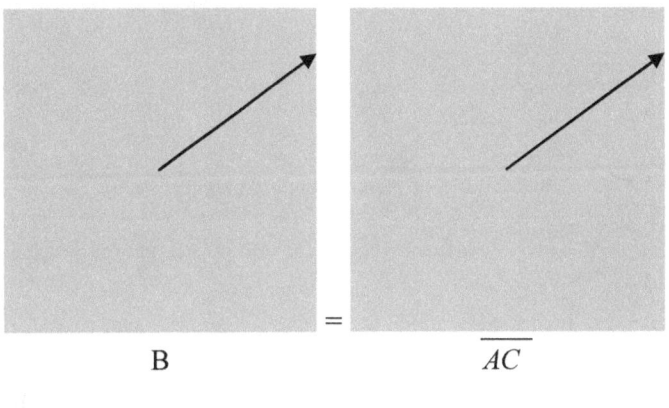

B = $\overline{AC}$

FIG. (12)

Every wave has three different properties: (1). wavelength (a spatial measurement), (2). frequency (a temporal measurement), and (3). energy (or equivalently mass). The exact meanings of the terms "space", "time" and "mass" still remain rather elusive, and while each of these wave properties can be measured, it seems that each can be measured only in terms of the other two. In fact, frequency, wavelength, and energy are simply three different ways of looking at the same wave. It is often instructive to look at something from a different angle, and looking at a wave from a different angle means looking at a wave "literally" from a different angle, a different phase angle.

Let the Planck waves $\vec{s} = \overrightarrow{\cos\theta_s} + \overrightarrow{i\sin\theta_s}, \vec{t} = \overrightarrow{\cos\theta_t} + \overrightarrow{i\sin\theta_t}$, and $\vec{m} = \overrightarrow{\cos\theta_m} + \overrightarrow{i\sin\theta_m}$ be three unit vectors corresponding to rotations in the complex plane, part real and part imaginary. $\vec{s}$ represents one Planck length, $\vec{t}$ represents one Planck time, and $\vec{m}$ represents one Planck mass. Again let their relative phase angles be

$\theta_s = 150°$
$\theta_t = 30°$
$\theta_m = 270°$

exactly as before. Let the three vectors now represent space, time and mass,

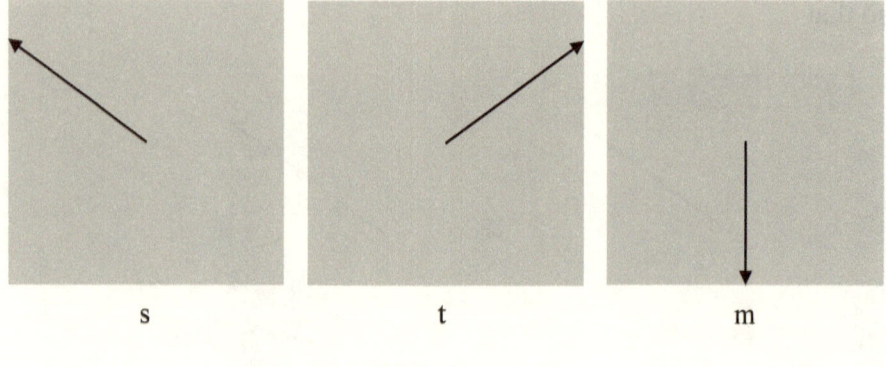

FIG. (13)

as opposed to negative space, negative time, and negative mass (i.e. the mass equivalent of negative energy), which of course would look like Fig. (14).

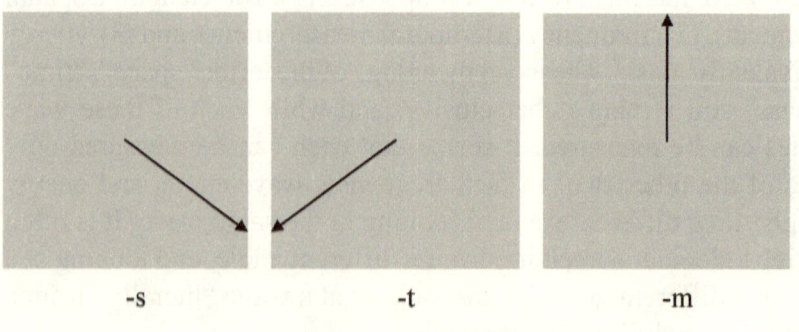

FIG. (14)

Again,

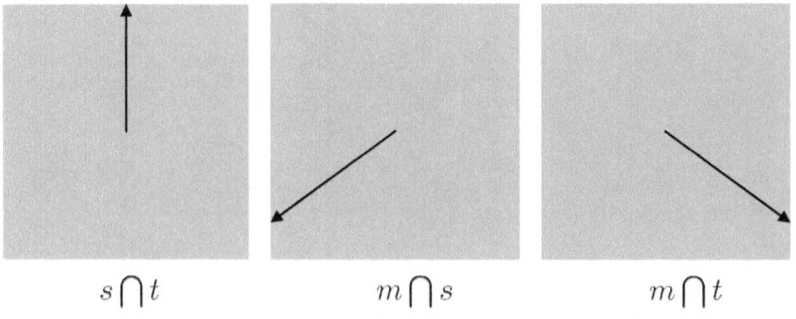

$$s \bigcap t \qquad m \bigcap s \qquad m \bigcap t$$

FIG. (15)

and

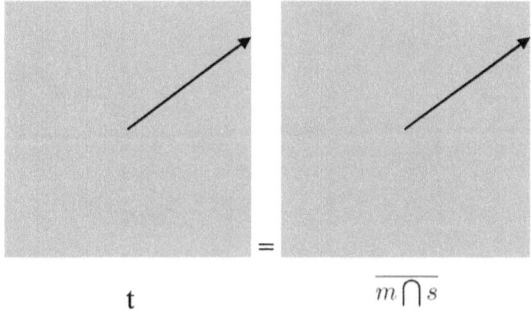

$$t \qquad = \qquad \overline{m \bigcap s}$$

FIG. (16)

and

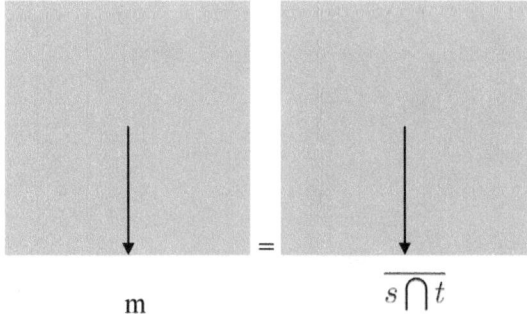

$$m \qquad = \qquad \overline{s \bigcap t}$$

FIG. (17)

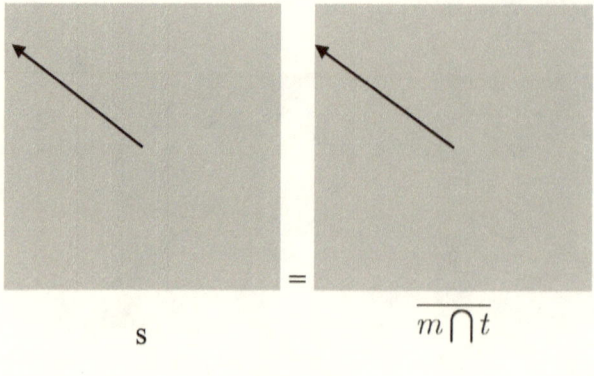

S $\overline{m \bigcap t}$

FIG. (18)

From this standpoint, it can be seen that the Planck wave itself

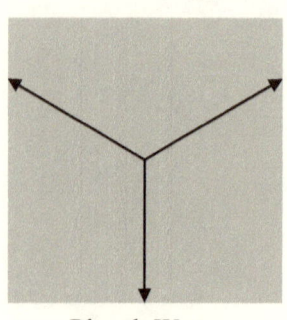

Planck Wave

FIG. (19)

is represented by the zero vector as in Fig. (20), since

$$\overrightarrow{(cos150 + isin150)} + \overrightarrow{(cos30 + isin30)} + \overrightarrow{(cos270 + isin270)} = \overrightarrow{0}.$$

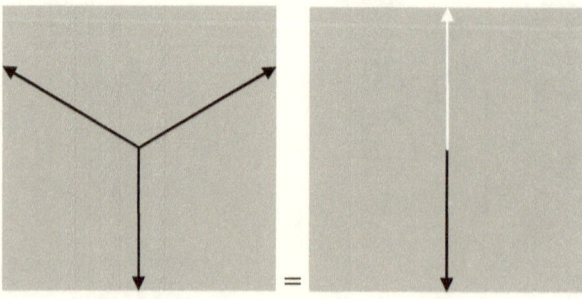

FIG. (20)

Here, $s \cap t$ is represented by the upward pointing white arrow, and mass is represented by the downward pointing black arrow.

The Planck wave is, in fact, the zero vector since, for any two Planck waves (wave$_1$) and (wave$_2$), although they have non zero energy (and consequently a non zero mass equivalent), their positive mass/energy is exactly negated by their negative gravitational potential energy.

$$m_1 c^2 - \frac{G(m_1)(m_2)^2}{r_p} = 0$$

and likewise

$$m_2 c^2 - \frac{G(m_1)(m_2)^2}{r_p} = 0$$

And it follows that, if the Planck wave represents the zero vector, then the relationship in Fig. (20) is equivalent.

The three representative vectors in the above discussion constitute what is called a linearly dependent set (as opposed to a linearly independent set). Technically, "a set with two or more vectors is linearly dependent if and only if at least one of the vectors is expressible as a linear combination of the other vectors in the set". The set as a whole is equal to the zero vector, in this case, the Planck wave. For instance, if

$$(\vec{0} = \vec{V_1} + \vec{V_2} + \vec{V_3})$$

for arbitrary vectors $\vec{V_1}$, $\vec{V_2}$, and $\vec{V_3}$, then

$$\vec{V_1} = (-1)(\vec{V_2}) + (-1)(\vec{V_3})$$

and

$$\vec{V_2} = (-1)(\vec{V_1}) + (-1)(\vec{V_3})$$

and

$$\vec{V_3} = (-1)(\vec{V_1}) + (-1)(\vec{V_2}).$$

The energy, or equivalently mass, of a wave is directly proportional to its frequency, whereas wavelength (the spatial measurement) and duration (the temporal measurement) are inversely proportional to frequency. Between any two arbitrary frequencies there exist an infinite number of intermediate frequencies. In terms of mass, space, and time therefore, an infinite spread of wave frequencies (representing unit vectors), where the highest frequency is infinitely higher than the Planck frequency, and lowest frequency is infinitely lower, would look something like Fig. (21), where mass is represented by a downward pointing black arrow and $s \bigcap t$ is represented by an upward pointing white arrow, as in Fig. (20).

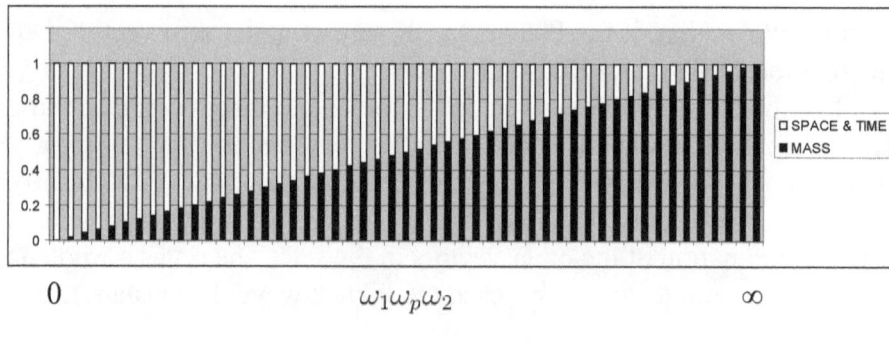

FIG. (21)

For convenience, let's number the waves to the left of the Planck wave (ω_p), with odd numbers, and waves to the right with even numbers (as shown above). Concentrating on just the three center frequencies it becomes apparent that the vectors

$$\vec{\omega_1} + \vec{\omega_2} = \vec{\omega_p}$$

In other words,

$$\frac{m_1 + m_2}{(s \cap t)_1 + (s \cap t)_2} = \frac{m_p}{(s \cap t)_p}$$

Likewise,

$$\vec{w_1} + \vec{w_2} + \vec{w_3} + \vec{w_4} = \vec{w_p}$$

and

$$\vec{w_1} + \vec{w_2} + \vec{w_3} + \vec{w_4} + - - - - - - + \vec{w}_\infty = \vec{w_p}$$

And since w_p constitutes the zero vector

$$\vec{0} = \vec{w_p} = \vec{w_1} + \vec{w_2} + \vec{w_3} + \vec{w_4} + - - - - - - + \vec{w}_\infty$$

is a dependent set. Every vector in the set is equal to a linear combination of the remaining vectors, but the set as a whole is equal to "nothing".

Much Ado about Nothing

Even an attempt to define "nothing" seems to present an irreconcilable paradox. From the *MERRIAM-WEBSTER DICTIONARY*:

> *NOTHING:* something that does not exist.

> *SOMETHING*: some indeterminate or unspecified thing.

> *THING*: a separate and distinct individual quality, fact, idea, or usually entity.

So that in summation,

> *NOTHING*: some indeterminate or unspecified separate and distinct individual quality, fact, idea, or usually entity that does not exist.

Every use of the word "nothing" seems to contain a paradox. "To be nothing." "To do nothing." "To know nothing." "To find nothing." "To see nothing." "To hear nothing." "To feel nothing."

The ancient Greeks, from whom we have inherited the foundations of western logic, seemed reticent to grapple with the paradoxes of nothingness. The Sicilian philosopher Empedocles dealt with the problem by maintaining that all of space was filled with a mysterious ether that guarded against a perfect vacuum; an idea, in fact, which persisted up until the beginning of the 20th century, when Einstein dispelled the idea of the luminiferous ether.

In fact the old concept of "nothing", as the existence of some location in space and time which is devoid of mass, is called into question by

Einstein's theory, since for a massless particle, space and time themselves cease to exist (or at least one dimension of three dimensional space, and the one dimension of time). Does this insight clarify the situation or simply deepen the paradox?

To illustrate the difficulty the human mind has with the concept of "nothing"; consider how long it took for modern humans to discover the number zero. Modern humans have remained largely unchanged for over 100,000 years. We know precious little about those early humans; however we must assume that they could count their fingers and toes. Even many animals have a rudimentary ability to count. Birds seem to remember the number of eggs in the nest. Dogs, cats, ducks, etc., keep track of the number of puppies, kittens, and ducklings for which they are responsible. And yet it was apparently not until sometime in the first half millennium AD (independently by the Babylonians, Mayans, and the peoples of the Indus Valley) that any civilization incorporated the zero into its counting system. This was more than 1500 years after the first recorded use of numbering systems by the Egyptians and Sumerians. Even the Greeks; Thales, Pythagoras, Euclid, Archimedes—the zero escaped them all.

In retrospect it would seem that mathematics could have progressed at a lightning pace if the zero had been the first number discovered. The highly successful arabic numeral system that we use today, which was borrowed from the Indian place value system, is based upon powers of ten since most humans learn to count with their fingers. With this system, it is possible to represent any positive integer, using some set of finger permutations. As we know, any number in this system is simply the summation of a set of numbers ($n = 0$ thru 9) which have been multiplied by some power of 10.

POWER	9	8	7	6	5	4	3	2	1	0
10^n	10^9	10^8	10^7	10^6	10^5	10^4	10^3	10^2	10^1	10^0
Times	x	x	x	x	x	x	x	x	x	x
n	n	n	n	n	n	n	n	n	n	n

For instance the integer 1,076,431,204 would be:

POWER	9	8	7	6	5	4	3	2	1	0
10^n	10^9	10^8	10^7	10^6	10^5	10^4	10^3	10^2	10^1	10^0
Times	x	x	x	x	x	x	x	x	x	x
n	1	0	7	6	4	3	1	2	0	4

In fact, just the organization of this table gives insight into another fascinating fact about the power of "nothing". The last entry on the right says that the integer

$$4 = 4 \text{ X } 10^0,$$

so that 10^0 must equal 1. In fact, as we know, any number raised to the zero power is equal to one. That is, any number multiplied by itself zero times is equal to 1. It's simple enough to illustrate. For instance take the fraction $\frac{A^x}{A^y}$. Algebra tells us that

$$\frac{A^x}{A^y} = A^{x-y}$$

so that if $y = x$,

$$\frac{A^x}{A^x} = 1 = A^{x-x} = A^0$$

The more you think about it, the harder it is to imagine how our ancestors got by for so long without "nothing".

While the true meaning of "nothing" seems to have frustrated even the Greeks, perhaps they were closer than they knew. In his first book of the elements Euclid tells us that "things that are equal to the same thing are equal to each other". Perhaps we should look more closely at those things which are equal to "nothing", which, by the way, would comprise a huge (perhaps infinite) list. For instance, every balanced equation obviously can be made equal to zero (and that's certainly a huge list). In fact, the solutions to equations are sometimes called the zeros of the equation.

To me, one of the most fascinating of these equations is the one known as Euler's identity, named for the Swiss mathematician Leonhard Euler. It says that

$$0 = e^{i\pi} + 1$$

The five most important constants in all of mathematics, all wrapped up in one simple equation. First, a little background on the derivation of the equation may be in order. The numbers e and π are members of the set called transcendental, so called because they *transcend* algebraic methods of determination. They can only be represented as the sums of infinite series. The number e for instance (which rounds to about 2.71828) is the sum of the infinite series

$$e = \frac{1}{0!} + \frac{1}{1!} + \frac{1}{2!} + \frac{1}{3!} + \frac{1}{4!} + \frac{1}{5!} \cdots \frac{1}{n!}$$

where n is a positive integer which approaches infinity and (!) signifies factorial.[1] By the way, zero factorial is equal to 1. "Ain't "nothing" cool".

The six basic trigonometric functions are transcendental functions, as are the inverse trigonometric functions, the exponential functions, and the logarithmic functions. It turns out that if you take the infinite series for the cosine of some angle (x) and add it to the infinite series for the imaginary sine of (x), the sum is equal to e^x. If we choose the angle (x) to be π (i.e. 180 °) then

$$cos\pi + i\sin\pi + 1 = e^{i\pi} + 1 = 0$$

hence Euler's Identity.

The point here is that if, as Euclid says, "things equal to the same thing are equal to each other" then "nothing", in this case, can be defined as an infinite series. For that matter, it can be defined as the superposition of four infinite series: $(\cos\pi)$, $(i\sin\pi)$, $(-\cos\pi)$, and $(-i\sin\pi)$. Each element of the series can be considered a vector, since each has both magnitude and direction (either + or -). Together this set of vectors constitutes a linearly dependent set. While the set as a whole is equal to zero, each individual vector is non zero, and in fact is equal to a linear combination of the remaining vectors.

Likewise, "nothing" can be defined as the sum of the infinite set of vectors to either side of the Planck wave, as shown in FIG. (21). In other words, if the Planck wave constitutes the zero vector, then the first wave to the left of the Planck wave in Fig. (21) is the negation of the first wave to the right, and consequently, the complete set to the left negates the complete set to the right. In the same way, the zero vector could be

---

[1] The factorial of a positive integer n, denoted by n!, is the product of all positive integers less than or equal to n.

represented by two or more sets of dependent sets, or an infinite number of sets of sets. It is fascinating how unconstrained one can be when talking about "nothing".

When closely scrutinized, "nothing" seems to spontaneously give up secrets. For instance, returning to Euler's identity,

$$0 = e^{i\pi} + 1$$

Twice as much "nothing" is still "nothing", so

$$0 + 0 = 2e^{i\pi} + 2$$
$$0 + 0 + 0 = 3e^{i\pi} + 3$$
$$0 + 0 + 0 = 3e^{i\pi} + 3$$
$$0 + 0 + 0 ----- = ne^{i\pi} + n$$
$$ne^{i\pi} = -n$$

Hence we can generate the infinite set of all positive integers and the infinite set of all negative integers. Together the two sets form the infinite set of all integers (Z). The ratios of the elements of (Z) of course form the infinite set of all rational numbers (Q). With (0, 1, e, i, π, Z, and Q) at our disposal we continue on.

$$e^{i\pi} = -1$$
$$\ln e^{i\pi} = \ln(-1)$$
$$i\pi = \ln(-1)$$
$$Qi\pi = Q\ln(-1)$$
$$Qi\pi = \ln(-1^Q)$$
$$e^{Qi\pi} = (-1^Q)$$
$$\cos(Q\pi) + i\sin(Q\pi) = (-1^Q)$$

or

$$0 = (cos(Q\pi) + i\sin(Q\pi)) - (-1^Q) \qquad (10.1)$$

where we now have cosine and sine functions for the continuous set of rotations (0°-360°) in the complex plane, which says, in this context, that "nothing" can be described as an infinite number of complex unit vectors.

All other trigonometric and inverse trigonometric functions can be derived from some combination of (cos θ) and (sin θ) where (θ = Qπ). In other words

$$\frac{\sin \theta}{\cos \theta} = \tan \theta$$

$$\frac{\cos \theta}{\sin \theta} = \cot \theta$$

$$(\cos \theta)^{-1} = \sec \theta$$

$$(\sin \theta)^{-1} = \csc \theta$$

arcsin $\theta = (\sin^{-1})\theta$, and arccos $\theta = (\cos^{-1})\theta$, etc. In fact:

$$\cos \theta = \frac{e^{i\theta} + e^{-i\theta}}{2}$$

$$\sin \theta = \frac{e^{i\theta} - e^{-i\theta}}{2i}$$

$$\cosh \theta = \frac{e^{\theta} + e^{-\theta}}{2}$$

$$\sinh \theta = \frac{e^{\theta} - e^{-\theta}}{2}$$

so that all trigonometric, inverse trigonometric, hyperbolic, and inverse hyperbolic functions can be generated.

Ivan Niven, a Canadian-American mathematician, published a proof in 1961 showing that, with the exception of the 60° angle, the cosine of a rational number given in degrees between 0° and 90° is irrational, so that we can generate an infinite set of irrational numbers in addition to e and π. All transcendental numbers are irrational and most irrational numbers are transcendental, so that out of an infinite set of irrationals we most likely can generate a sizeable set of transcendentals.

From "nothing" therefore we have generated the five most important constants in all of mathematics (actually Euler did), the infinite set of integers and rational numbers, the infinite set of imaginary numbers, an infinite set of irrational numbers, a sizable set of transcendental numbers (both real and imaginary in each case); and the complete set of trigonometric, inverse trigonometric, hyperbolic, and inverse hyperbolic functions. The possibilities seem to be limitless when you have "nothing" to work with. Fascinating!

Quantum Gravity

Imagine the infinite set of waves, as in Fig. (22), to either side of an arbitrary frequency ω_0, where the black portion represents the mass vector and the white portion represents the $s \cap t$ vector; the black vector pointing down and the white vector pointing up. Label three frequencies ω_1, ω_0 and ω_2 from left to right, respectively, as shown.

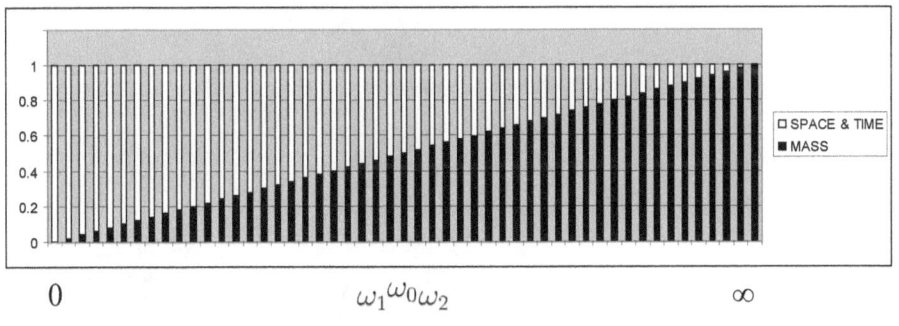

FIG (22)

$$\frac{\omega_1}{k_1} = \frac{\omega_0}{k_0} = \frac{\omega_2}{k_2} = c \tag{11.1}$$

$$\frac{d\omega_1}{dk_2} = v = \left(\frac{\omega_1}{\omega_2}\right)c \tag{11.2}$$

$$\left.\begin{array}{c}\\ \\ \end{array}\right\} uv = c^2, \quad \frac{v}{u} = \left(\frac{\omega_1}{\omega_2}\right)^2 = \frac{v^2}{c^2} \tag{11.3}$$

$$\left.\begin{array}{c}\\ \\ \end{array}\right\} uv = c^2, \quad \frac{u}{v} = \left(\frac{\omega_2}{\omega_1}\right)^2 = \frac{u^2}{c^2} \tag{11.4}$$

$$\frac{d\omega_2}{dk_1} = u = \left(\frac{\omega_2}{\omega_1}\right)c \tag{11.5}$$

In view of Eq. (11.3) the gamma factor now becomes

$$\frac{1}{\sqrt{1 - \frac{\omega_1^2}{\omega_2^2}}} \tag{11.6}$$

and since

$$1 - \left(\frac{\omega_1}{\omega_2}\right)^2 = \left(\frac{\omega_2^2 - \omega_1^2}{\omega_2^2}\right) = \frac{(\omega^2)_{dif}}{(\omega_2^2)}$$

(where $(\omega^2)_{dif}$ is the difference in the squares of ω_1 and ω_2, and ω_{dif} would be the square root of that difference). Consequently, the equations of special relativity become

$$\frac{t}{t_0} = \frac{\omega_{dif}}{\omega_2} \tag{11.7}$$

$$\frac{r}{r_0} = \frac{\omega_{dif}}{\omega_2} \tag{11.8}$$

and

$$\frac{m_0}{m} = \frac{\omega_{dif}}{\omega_2} \tag{11.9}$$

and we can see that, at the fundamental level, intelligent observers are replaced by simple waves, and what were, at first, considered to be intelligent observations, come about quite naturally. Nature seems to have an intelligence all its own.

Now since wavelength varies inversely with frequency (i.e. wave number varies directly with frequency), it follows that for any two arbitrary frequencies, ω_1 and ω_2,

$$\frac{\omega_1}{r_2} = \frac{\omega_2}{r_1} = k, \text{ a constant}$$

For instance, for arbitrary frequencies 4 sec^{-1}, and 6 sec^{-1}

$$\frac{\omega_6}{r_4} = \frac{\frac{6}{4}\omega_4}{r_4} = \frac{\omega_4}{\frac{4}{6}r_4} = \frac{\omega_4}{r_6}$$

Therefore, for any two arbitrary frequencies (ω_1 and ω_2), there exists an intermediate frequency (ω_0), such that $\omega_2 = \Omega\omega_0$ and $\omega_1 = \frac{\omega_0}{\Omega}$ for some scalar constant Ω. Energy, or equivalently mass, varies directly with frequency so that, in contrast to Eq. (11.1)

$$\frac{dm_1}{dr_2} = \frac{m_0}{r_0} = \frac{dm_2}{dr_1} = k, \text{ a constant} \tag{11.10}$$

where

$dm_1 = m_0 - m_1$
$dm_2 = m_2 - m_0$
$dr_1 = r_1 - r_0$
$dr_2 = r_0 - r_2$

If we make ω_0 the Planck frequency substituting $\frac{m_p}{r_p}$ for $\frac{m_0}{r_0}$, then the constant becomes

$$\frac{m_p}{r_p} = \frac{2.18X10^{-08}kg}{1.62X10^{-35}m} = 1.35X10^{27}kg/m = \frac{c^2}{G} \qquad (11.11)$$

and Eq. (11.10) becomes

$$\frac{dm_1}{dr_2} = \frac{m_p}{r_p} = \frac{dm_2}{dr_1} = \frac{c^2}{G} \qquad (11.12)$$

and the spread of frequencies looks like Fig. (23)

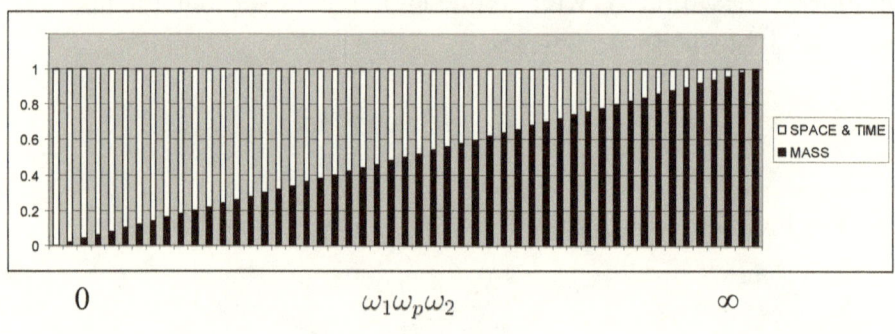

Fig. (23)

where the highest frequency is infinitely higher than the Planck frequency and the lowest frequency is infinitely lower.

Just as the change in space with respect to the change in time is a constant as shown in Eq. (11.1), Eq. (11.10) says that the change in mass with respect to the change in space is also a constant.

The Planck wave constitutes two equal and opposite vectors, i.e. the zero vector, as shown in FIG. (20). There is perfect symmetry between:

1. Left-handed space and right-handed space
2. forward time and reverse time
3. Positive energy and negative energy (i.e. Positive mass and negative mass)

Assuming that positive mass is the negation of $s \cap t$, waves to the left of ω_p constitute an asymmetry of $s \cap t$ over mass (or equivalently negative mass over positive mass). Waves to the right of ω_p constitute an asymmetry of mass over $s \cap t$ (or equivalently positive mass over negative mass).[2]

Concentrating on ω_1, ω_p, and ω_2, and remembering that the vector sum of the three vectors is still just the zero vector,

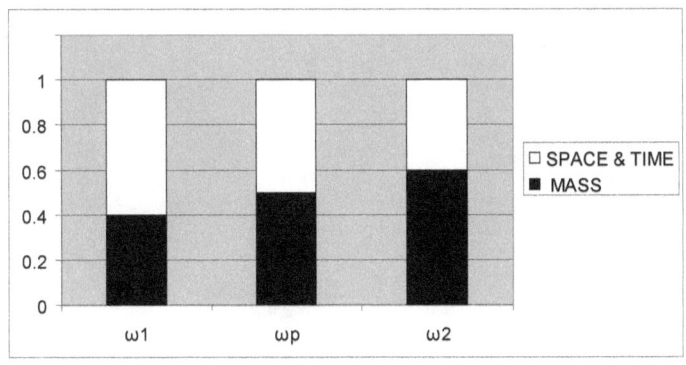

Fig. (24)

Eq. (11.10) again is

$$\frac{dm_1}{dr_2} = \frac{m_p}{r_p} = \frac{dm_2}{dr_1} = \frac{c^2}{G} \qquad (11.13)$$

so that in contrast to Eqs. (11.2)-(11.5) we have

$$\frac{dm_1}{dr_1} = \left(\frac{\omega_1}{\omega_p}\right)^2 \left(\frac{m_p}{r_p}\right) \qquad (11.14)$$

$$\left.\right\} \frac{\frac{dm_1}{dr_1}}{\frac{m_p}{r_p}} = (\frac{\omega_1}{\omega_p})^2 = \frac{v^2}{c^2} \quad (11.15)$$

$$\left.\right\} \frac{\frac{dm_2}{dr_2}}{\frac{m_p}{r_p}} = (\frac{\omega_2}{\omega_p})^2 = \frac{u^2}{c^2} \quad (11.16)$$

$$\frac{dm_2}{dr_2} = \left(\frac{\omega_2}{\omega_p}\right)^2 \left(\frac{m_p}{r_p}\right) \qquad (11.17)$$

---

[2] For simplicity of discussion, references herein will refer to left handed space as positive space, forward moving time as positive time, and Newtonian mass as positive mass.

Eqs. (11.2)-(11.5) determine the quantum laws of special relativity while Eqs. (11.14)-(11.17) determine the quantum laws of general relativity. From Eq. (11.15) and Eq. (11.16) we see that the gamma function becomes either

$$\gamma = \frac{1}{\sqrt{1 - \frac{\frac{dm_1}{dr_1}}{\frac{m_p}{r_p}}}} \tag{11.18}$$

or

$$\gamma = \frac{1}{\sqrt{1 - \frac{\frac{dm_2}{dr_2}}{\frac{m_p}{r_p}}}} \tag{11.19}$$

and it follows that

$$\gamma_1 = \frac{1}{\sqrt{1 - \left(\frac{\omega_1}{\omega_p}\right)^2}} \tag{11.20}$$

and

$$\gamma_2 = \frac{1}{\sqrt{1 - \left(\frac{\omega_2}{\omega_p}\right)^2}} \tag{11.21}$$

and of course as $(\omega_1 \rightarrow \omega_p \leftarrow \omega_2)$, time slows to a halt, and spatial dimensions shrink (the beginning of time and space). At that point, the gravitational force, and what will come to be known as the Planck force, are equal and opposite (the zero vector). The two forces will diverge, however, into two separate forces.

These quantum equations determine the laws of general relativity and space-time curvature. This curvature of space-time is the direct result of the presence of mass, and it is this curvature which we perceive as gravitation, and, after all, the general theory of relativity is, in fact, a theory of gravity; quantum gravity, in this case.

If we look closely at Eq. (11.18), we recall that the factor $(\frac{m_p}{r_p})$ is equal to $(\frac{c^2}{G})$, so that the equation becomes

$$\gamma = \frac{m}{m_0} = \frac{1}{\sqrt{1 - \left(\frac{Gm_0}{rc^2}\right)^2}} \qquad (11.22)$$

or

$$\frac{m_0^2}{m^2} = 1 - \frac{(m_0)(G)}{(r)(c^2)}$$

so that

$$\frac{m_0^2}{m^2} = \frac{(c^2)(r) - (Gm_0)}{(c^2)(r)}$$

and

$$m_0 c^2 = mc^2 - \frac{Gmm_0}{r}$$

or

$$\frac{Gmm_0}{r} = mc^2 - m_0 c^2 \qquad (11.23)$$

For example, the sun is the largest mass in our neighborhood, being about $m_s = 2.00 \times 10^{30} kg$. At the point where we on the earth are located (about $r = 1.51 \times 10^{11}$ meters away from the sun) we would, according to Eq. (11.18), experience a gamma factor (γ) of about

$$\gamma = \frac{1}{\sqrt{1 - \frac{\frac{2.00 X 10^{30} kg}{1.51 X 10^{11} m}}{\frac{2.17 X 10^{-8} kg}{1.62 X 10^{-35} m}}}} \approx 1.000000005$$

$$\gamma = \frac{m}{m_0}$$

so that

$$m = \gamma m_0$$

And for any arbitrary mass (m_0) located at that point (such as you, or I, or the planet earth), the difference in mass/energy

$$mc^2 - m_0 c^2$$

as shown in Eq. (11.23), would represent the kinetic energy of that arbitrary mass. Therefore,

$$\gamma m_0 c^2 - m_0 c^2 = \frac{1}{2} m_0 v^2 = \frac{1}{2} m_0 \frac{(2\pi r)^2}{t^2} \qquad (11.24)$$

where ($2\pi r$) is the earth's orbital distance around the sun and t is the orbital period.

The arbitrary mass cancels out, so that in solving for t we have

$$t = \sqrt{\frac{2(\pi r)^2}{(\gamma c^2 - c^2)}} \qquad (11.25)$$

Plugging in the numbers, we arrive at a result which agrees rather well with that of our Neolithic ancestors:

$$t = \sqrt{\frac{2(3.14)^2 (1.51 X 10^{11} m)^2}{(\gamma c^2) - (c^2)}} = 3.16 X 10^7 sec \approx 365 days \qquad (11.26)$$

At the same time, since

$$(mc^2 - m_0c^2) = \frac{1}{2}m_0v^2$$

it follows from (11.23) that

$$\frac{Gmm_0}{r} = \frac{1}{2}m_0v^2$$

and

$$v^2 = \frac{2Gm}{r}$$

so that

$$t = t_0\sqrt{1 - \frac{2Gm}{rc^2}}$$

which agrees with Eq. (3.1).

We will return shortly and discuss all this in further detail. For now, let us attack this from the opposite direction.

Mass swallows up space and time. A large mass warps the space around it and causes a clock near it to run more slowly than an identical clock in empty space. The direct effect of mass on space and time can be derived from the gravitational constant itself.

$$G = 6.67 X 10^{-11} Nm^2/kg^2$$
$$= 6.67 X 10^{-11}(kgm/sec^2)(m^2/kg^2)$$
$$= (6.67 X 10^{-11}m^3/sec^2)/kg \qquad (11.27)$$

or $6.67 X 10^{-11}$ cubic meters per second per second per kilogram. This can be interpreted to mean that if 1 kg of mass were to disappear from our universe, the rate of expansion of the universe would consequently increase by 6.67 x 10^{-11} cubic meters per second per second. Inversely, if the amount of mass in our universe were to increase by 1kg, the rate of expansion would decrease by a like amount.

It is more convenient for the present to speak in terms of G^{-1} (G inverse) i.e. $\frac{1}{G}$.

$$G^{-1} = 1.5 X 10^{10} kg/(m^3/sec^2) \tag{11.28}$$

If we want to express this in units of energy rather than units of mass, we do so by multiplying by the square of the speed of light, which gives

$$G^{-1}c^2 = 1.35 X 10^{27} J/(m^3/sec^2)$$

But

$$(G^{-1})(c^2) = 1.35 X 10^{27} J/(m^3/sec^2)$$
$$= 1.35 X 10^{27} ((kg)(m^2)/sec^2)/(m^3/sec^2)$$
$$= 1.35 X 10^{27} kg/m \tag{11.29}$$

which is Eq. (11.12); i.e.:

$$\frac{c^2}{G} = \frac{m_p}{r_p} \tag{11.30}$$

or inversely

$$\frac{r_p}{m_p} = \frac{G}{c^2} \tag{11.31}$$

Think of it this way. The rate of change in space with respect to the change in time (the speed of light) is a constant (299,792,458 m/sec). Eq. (11.29) says that the rate of change in mass with respect to the change in space is a constant ($1.35 X 10^{27} kg/m$). The product of the two, of course, would be the rate of change in mass with respect to the change in time, which would be

$$\frac{c^3}{G} = 4.04 X 10^{35} kg/sec$$

If we measure distance in light seconds and make $(4.04 X 10^{35} kg)$ a unit of mass (call it a Planckogram for lack of a better term), then one Planckogram of mass is equivalent to one second of time and one light second of space, and the vector sum of the three, as in Fig. (20), is the zero vector.

Multiplying both sides of Eq. (11.31) by $\frac{dm}{(dr)^2}$ we have

$$\frac{(dm)(r_p)}{(dr^2)(m_p)} = \frac{(dm)(G)}{(dr)^2(c^2)}$$

The mass dm can be expressed in units of energy by multiplying both sides by the square of the speed of light

$$\frac{(dm)(r_p)(c^2)}{(dr^2)(m_p)} = \frac{(dm)(G)}{(dr)^2}$$

or

$$\left(\frac{\frac{dm}{dr}}{\frac{m_p}{r_p}}\right)\left(\frac{c^2}{dr}\right) = \frac{(dm)(G)}{(dr)^2} \tag{11.32}$$

which is, of course, the gravitational acceleration resulting from a mass dm at a distance dr. For instance, if we were to replace dm with the earth's mass and dr with the earth's radius then

$$\left(\frac{\frac{5.97 X 10^{24} kg}{6.37 X 10^6 m}}{\frac{2.18 X 10^{-8} kg}{1.62 X 10^{-35} m}}\right)\left(\frac{(3.0 X 10^8 m/sec)^2}{6.37 X 10^6 m}\right) = \frac{(5.97 X 10^{24}) kg)(6.67 X 10^{-11} N m^2/kg^2)}{(6.37 X 10^6 m)^2}$$

$$= 9.8 m/sec^2$$

which is, of course, the gravitational acceleration at the earth's surface.

From Eq. (11.32), the gravitational acceleration

$$a = \frac{(dm)(G)}{(dr^2} = \frac{v^2}{dr}$$

so that

$$\left(\frac{\frac{dm}{dr}}{\frac{m_p}{r_p}}\right)\left(\frac{c^2}{dr}\right) = \frac{v^2}{dr} \tag{11.33}$$

or

$$\frac{\frac{dm}{dr}}{\frac{m_p}{r_p}} = \frac{v^2}{c^2} \tag{11.34}$$

Recalling the gamma factor from the theory of relativity

$$\gamma = \frac{1}{\sqrt{1 - \frac{v^2}{c^2}}}$$

and making the above substitution

$$\gamma = \frac{1}{\sqrt{1 - \frac{\frac{dm}{dr}}{\frac{m_p}{r_p}}}} \tag{11.35}$$

which agrees with Eq. (11.18).

Eq. (11.32) and the example above show that every (dm) and every (dr) in the universe are measured with respect to the Planck mass and Planck distance (base Planck units). The same is true with respect to (dt) and the Planck time. And in fact, the derivative $(\frac{dm}{dr})$ in Eq. (11.32) will coincide precisely with one of the waves in Fig. (23).

For instance, $(\frac{dm}{dr})$ for the earth, in the above example, would be about $9.37 X 10^{17} kg/m$. If the Planck wave has a value of $1.35 X 10^{27} kg/m$, then somewhere to the left of the Planck wave in Fig. (23), there exists a wave such that $(\frac{h}{r^2 c}) = 9.37 X 10^{17} kg/m$. In fact it would coincide with a wavelength of about $3.85 X 10^{-30} m$.

While "velocity" $(\frac{d\omega}{dk})$, or equivalently $(\frac{dr}{dt})$ is a familiar concept, $(\frac{dm}{dr})$ is not, and consequently we have no word for it. In three dimensions it is roughly equivalent to the rate of change in mass density. Perhaps "viscosity" would be an appropriate word, since the universal soup gets thicker as $\frac{dm}{dr}$ increases. In any event, just as[3]

$$\frac{v^2}{c^2} = \frac{v}{u} = \frac{\frac{d\omega}{dk}}{\int \frac{d\omega}{dk}}$$

where $\frac{\int d\omega}{\int dk}$ represents the set of all $\frac{d\omega}{dk}$'s, it can be seen that, in Eq. (11.32)

$$\frac{\frac{dm}{dr}}{\frac{m_p}{r_p}} = \frac{v^2}{c^2} = \frac{v}{u} = \frac{\frac{d\omega}{dk}}{\frac{\int d\omega}{\int dk}}$$

And the Planck wave vector in Fig. (23) is, in fact, equal to a linear combination of all the remaining vectors in the set, which begs the question again, "does entropy have something to do with relativity?". Does the set of all $\frac{d\omega}{dk}$'s represent the number of available microstates? After all, according to Eq. (11.18) and Eq. (11.35), time slows down as

$$\left(\frac{dm}{dr}\right) \rightarrow \left(\frac{m_p}{r_p}\right)$$

and speeds up as

$$\left(\frac{dm}{dr}\right) \rightarrow 0$$

Again, we will return shortly to discuss Eq. (11.18) and Eq. (11.35).

Returning now to Eq. (11.12) and expressing it in units of energy by multiplying by the square of the speed of light,

---

[3]   Eq. (4.5)

$$\frac{d(m_1c^2)}{dr_2} = \frac{(m_pc^2)}{r_p} = \frac{d(m_2c^2)}{dr_1} = \frac{c^4}{G} \qquad (11.36)$$

The rate at which energy changes with respect to distance is called "force", and Eq. (11.36) says that the rate of change in the energy of wave$_1$ $(\hbar\omega_p - \hbar\omega_1)$ with respect to the change in the radial length of wave$_2$ $(\frac{1}{k_p} - \frac{1}{k_2})$ is a constant force which we know as the Planck force.

The Planck force is a tremendous force with a constant strength of about $1.21 X 10^{44} newtons$. From Eq. (11.12)

$$\frac{dm_2}{dr_1}G = c^2 \qquad (11.37)$$

so that

$$\frac{dm_1}{dr_1}\frac{dm_2}{dr_1}G = \frac{dm_1}{dr_1}c^2 \qquad (11.38)$$

Or

$$\frac{d(m_1c^2)}{dr_1} = \frac{dm_1 dm_2}{dr_1^2}G \qquad (11.39)$$

And likewise

$$\frac{d(m_2c^2)}{dr_2} = \frac{dm_1 dm_2}{dr_2^2}G \qquad (11.40)$$

Eq. (11.39) says that the rate of change in the energy of a wave $(m_p - m_1)c^2$ with respect to the change in its own radial length $(\frac{1}{k_1} - \frac{1}{k_p})$ is equal to a force between dm_1 and dm_2 which varies directly with the product of the masses and inversely with the square of dr_1, and the right side of the equation identifies that force as Newton's gravitational force. Eq. (11.40) says the same thing with respect to dr_2, keeping in mind that dr_1 and dr_2 are on opposite sides of the Planck barrier. At time zero, the Planck force vector and the gravitational force vector are equal and opposite (the zero vector). This situation is unstable, however, and this symmetry will become broken.

From Eq. (11.39), if

$$\frac{d(m_1c^2)}{dr_1} = \frac{dm_1dm_2}{dr_1^2}G$$

then

$$d(m_1c^2) = \frac{dm_1dm_2}{dr_1}G$$

The right-hand side of the equation is gravitational potential energy, so that

$$0 = d(m_1c^2) - \frac{dm_1dm_2}{dr_1}G \qquad (11.41)$$

and likewise

$$0 = d(m_2c^2) - \frac{dm_1dm_2}{dr_2}G \qquad (11.42)$$

This says that the negative energy of the gravitational potential exactly negates the positive energy of mass, *just as Peter Bergmann hypothesized.*[4]

Eqs. (11.36) through (11.40) follow quite naturally in that since, for any dm_1, dm_2, dr_1, and dr_2 the products of the magnitudes of the Planck force and the gravitational force are as follows

$$\left|\left(\frac{dm_1dm_2}{dr_1^2}G\right)\right|\left|\left(\frac{c^4}{G}\right)\right| = \left|\left(\frac{dm_1c^2}{dr_1}\right)\right|\left|\left(\frac{dm_2c^2}{dr_1}\right)\right| \qquad (11.43)$$

and

---

[4]   Page 56

$$\left|\left(\frac{dm_1 dm_2}{dr_2^2}G\right)\right|\left|\left(\frac{c^4}{G}\right)\right| = \left|\left(\frac{dm_1 c^2}{dr_2}\right)\right|\left|\left(\frac{dm_2 c^2}{dr_2}\right)\right| \quad (11.44)$$

"May The Force be with you" (the Planck force that is). In fact the Planck force is with us constantly, though we fail to recognize it. For instance from Eq. (11.43)

$$\left|\left(\frac{dm_1 c^2}{dr_1}\right)\right|\left|\left(\frac{dm_2 c^2}{dr_1}\right)\right| = \left|\left(\frac{dm_1 uv}{dr_1}\right)\right|\left|\left(\frac{dm_2 uv}{dr_1}\right)\right|$$

$$= \left|\left(\frac{dm_1 u^2}{dr_1}\right)\right|\left|\left(\frac{dm_2 v^2}{dr_1}\right)\right|$$

so that Eq. (11.43) becomes

$$\left|\left(\frac{dm_1 dm_2}{dr_1^2}G\right)\right|\left|\left(\frac{c^4}{G}\right)\right| = \left|\left(\frac{dm_1 u^2}{dr_1}\right)\right|\left|\left(\frac{dm_2 v^2}{dr_1}\right)\right| \quad (11.45)$$

Suppose we now replace dm_1 and dm_2 with, for instance, the sun's mass and the earth's mass, and dr_1 with the earth's orbital radius; in which case we now have

$$\left|\left(\frac{m_s m_e}{r^2}G\right)\right|\left|\left(\frac{c^4}{G}\right)\right| = \left|\left(\frac{m_s u^2}{r}\right)\right|\left|\left(\frac{m_e v^2}{r}\right)\right|$$

If we make $\frac{m_e v^2}{r}$ equal to the gravitational force between the earth and sun, the equation of course becomes simply

$$\left|\left(\frac{c^4}{G}\right)\right| = \left|\left(\frac{m_s u^2}{r}\right)\right|$$

The earth's orbital velocity (v) around the sun is about $29,700$ meters per second and its orbital radius is about $1.51 X 10^{11}$ meters, while the sun's mass is about $2X10^{30}$ kg. $u = \frac{c^2}{v} = 3.02X10^{12}$ meters per second. Plugging in the numbers we see that

$$\frac{m_s u^2}{r} = \frac{(2.00X10^{30}kg)(3.02X10^{12}m/sec)^2}{1.51X10^{11}m} = 1.21X10^{44}newtons$$

which is, of course, the value of the Planck force.

For that matter, imagine Isaac Newton standing on the surface of the earth, and replace dm_1, dm_2, and dr_1 with the earth's mass, Isaac's mass, and the earth's radius, respectively. Whether it's Isaac, the queen of England, or an African elephant, each would feel a gravitational acceleration of 9.8 meters per second per second. If

$$\frac{v^2}{r_e} = 9.8m/sec^2$$

Then

$$u^2 = \frac{c^4}{v^2} = \frac{c^4}{(9.8m/sec^2)(r_e)}$$

and

$$\frac{(m_e)(u^2)}{r_e} = \frac{(5.97X10^{24}kg)(8.08X10^{33}m^4/sec^4)}{(9.8m/sec^2)(6.38X10^6m)^2}$$

$$= 1.21X10^{44}newtons$$

again, exactly equal to the Planck force. (I think I can feel it now. Fascinating!)

Returning now to Eq. (11.39)

$$\frac{dm_1}{dr_1}c^2 = \frac{dm_1 dm_2}{dr_1^2}G = \left(\frac{dm_1}{dr_1}\right)\left(\frac{dm_2}{dr_1}\right)G \qquad (11.46)$$

And from Eq. (11.12)

$$\frac{dm_2}{dr_1} = \frac{m_p}{r_p}$$

so that

$$\frac{dm_1}{dr_1}c^2 = \left(\frac{dm_1}{dr_1}\right)\left(\frac{m_p}{r_p}\right)G \qquad (11.47)$$

and

$$\left(\frac{\frac{dm_1}{dr_1}}{\frac{m_p}{r_p}}\right)\left(\frac{c^2}{dr_1}\right) = \frac{(dm_1)(G)}{(dr_1)^2} \qquad (11.48)$$

which agrees with Eq. (11.32).

Continuing on, just as waves can superimpose to behave as particles, particles can superimpose to behave as waves; even on the largest possible scale. Returning to Eq. (11.29), and the significance thereof, if we contend that the Planck wave represents the zero vector, then three times zero is still zero, in fact a billion, or trillion, or any number times zero is still zero. Fig. (25) shows three Planck wave vectors representing "nothing".

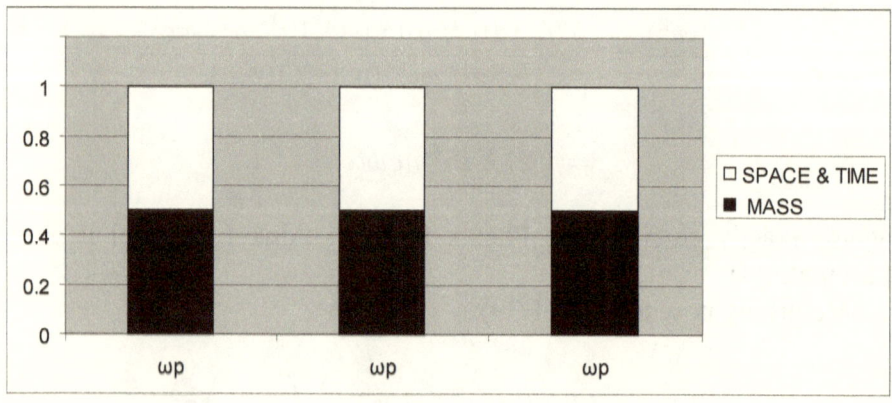

Fig. (25)

Now suppose that we perform a mind experiment. Let's take a small portion of mass from the left-hand vector and move it to the right-hand vector, as in (Fig. (26).

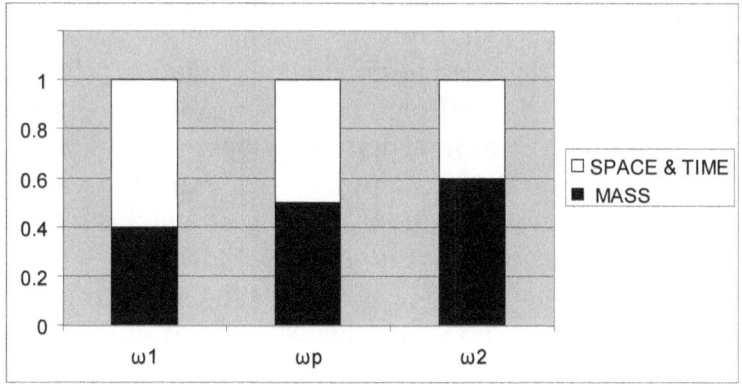

Fig. (26)

which is just Fig. (24) again. For that matter, moving mass from the left-hand vector to the right-hand vector, is equivalent to moving $s \cap t$ from the right-hand vector to the left-hand vector.

Now according to Eqs.(11.18) and(11.19), and including the gamma function for $\frac{m_p}{r_p}$

$$\gamma_1 = \frac{1}{\sqrt{1 - \frac{\frac{dm_1}{dr_1}}{\frac{m_p}{r_p}}}} \quad \gamma_p = \frac{1}{\sqrt{1 - \frac{\frac{m_p}{r_p}}{\frac{m_p}{r_p}}}} \quad \gamma_2 = \frac{1}{\sqrt{1 - \frac{\frac{dm_2}{dr_2}}{\frac{m_p}{r_p}}}} \tag{11.49}$$

$$\frac{t_1}{t_0} = \sqrt{1 - \frac{\frac{dm_1}{dr_1}}{\frac{m_p}{r_p}}} \quad \frac{t_p}{t_0} = \sqrt{1 - \frac{\frac{m_p}{r_p}}{\frac{m_p}{r_p}}} \quad \frac{t_2}{t_0} = \sqrt{1 - \frac{\frac{dm_2}{dr_2}}{\frac{m_p}{r_p}}} \tag{11.50}$$

$$\frac{r_1}{r_0} = \sqrt{1 - \frac{\frac{dm_1}{dr_1}}{\frac{m_p}{r_p}}} \quad \frac{r_p}{r_0} = \sqrt{1 - \frac{\frac{m_p}{r_p}}{\frac{m_p}{r_p}}} \quad \frac{r_2}{r_0} = \sqrt{1 - \frac{\frac{dm_2}{dr_2}}{\frac{m_p}{r_p}}} \tag{11.51}$$

γ_1 represents our own portion of the universe where space, time, and mass are real; γ_p is an event horizon where time stands still and one spatial dimension disappears; and γ_2 is a black hole where, according to the equations above, space, time, and mass become imaginary. In our mind experiment we moved mass from the real universe, across an event horizon and into a black hole, and Fig. (26) is a vector representation of those three separate partitions, and each of the vectors corresponds to a particular wave in Fig. (23).

In fact, the most fascinating thing about equation (11.49) is that the real and imaginary partitions evolve in exactly the same way. Real $s \cap t$ approaches 1 as imaginary $s \cap t$ approaches i. As positive mass is removed from the real universe, that real universe expands. As positive mass moves into the black hole, it manifests itself there as negative $s \cap t$ (imaginary negative $s \cap t$). In fact, the Schwarzchild radius of a black hole increases in direct proportion to the amount of mass that has fallen into it, and the constant of proportionality is (guess what),

$$\frac{2G}{c^2} = 2\left(\frac{r_p}{m_p}\right)$$

Remember, moving mass from the left-hand vector in Fig. (25) to the right-hand vector was equivalent to moving $s \cap t$ from the right-hand vector to the left-hand vector. Moving positive mass across the event horizon from the real universe to the imaginary universe, where it is manifested as negative $s \cap t$, is equivalent to moving negative mass across the event horizon from the imaginary universe to the real universe, where it is manifested as positive $s \cap t$. Remember how "fast cars and fine clothes" were manifested as "negative dollars" and vice versa?

I like to compare the scenario to hole flow in a semiconductor. Semiconductor materials consist of crystal lattices of positively charged atoms. These positive charges are called holes. While holes don't actually move, from the standpoint of relative motion an electron moving, say from left to right, is equivalent to a hole moving from right to left. Semiconductor devices such as diodes consist of one-way barriers called PN junctions, which allow electron currents to flow in one direction, but not the other. Of course, electron flow from left to right across the barrier is equivalent to hole flow from right to left across the barrier. The event horizon of a black hole is also a one-way barrier. Positive mass goes in but can't come out. Positive $s \cap t$ comes out, but it can't go in.

As with any exchange, there is an official exchange rate, and the exchange rate for mass with respect to space is $1.35 X 10^{27} kg/m$, and the exchange rate for mass with respect to time is $4.04 X 10^{35} kg/sec$. Of course, if space is measured in light seconds, and mass is measured in "Planckograms" the exchange rate is $1 pg/lightsec$, and $1 pg/sec$.

And in agreement with Eq. (11.36), the rate of change in real energy $(d(m_1 c^2))$, with respect to the change in imaginary space (dr_2), is equal to the rate of change in imaginary energy $(d(m_2 c^2))$, with respect to the change in real space (dr_1), both of which are equal to the Plank force $(\frac{m_p c^2}{r_p} = \frac{c^4}{G})$, *the force driving the universal expansion.*

At the same time, and in agreement with Eqs. (11.39) and (11.40), the rate of change in real energy $(d(m_1 c^2))$, with respect to the change in real space (dr_1), is equal to the rate of change in imaginary energy $(d(m_2 c^2))$, with respect to the change in imaginary space (dr_2), both of which are equal to a gravitational force between the real and imaginary universes, a force which is futilely attempting to resist the universal expansions (both real and imaginary) brought about by the Planck forces. Note, by the way, that this gravitational attraction extends only up to the event horizon in either case (Eqs. (11.39) and (11.40)).

While there are probably billions of black holes in the universe, $\gamma_{2total} = iQ_1 + iQ_2 + iQ_3 .... + iQ_n$ for rational numbers Q_n (less than 1). γ_{2total} approaches i as γ_1 approaches 1. Like pages in a book, every black hole reflects a segment of the history of the real universe; a mirror image, where positive mass is replaced by negative mass, and time runs backward (CPT theorem—?). As the amount of positive mass in the real universe decreases, the mutual gravitational force between its constituent masses decreases, but the Planck force remains constant $(\frac{c^4}{G})$, so that the expansion of the universe accelerates at the rate of 6.67 x 10^{-11} cubic meters per second per second per kilogram. And just as the energy driving Frank's fleet of trucks was equal to the Frank force times the distance driven, the energy driving this universal expansion is equal to the Planck force times the distance driven, i.e.,

$$(dm)c^2 = \left(\frac{c^4}{G}\right)(dr_U) \qquad (11.52)$$

where dr_U is the total change in the universal radius at that point in time.

The rate of expansion of the universe at any particular point in time is called the Hubble constant, and it is measured in kilometers per second per megaparsec, and its present value is about 71. This rate is not actually a constant however, as was once believed. The rate of expansion is constantly increasing, with this expansion being attributed to what is called dark energy (i.e. Eq. (11.52)). A megaparsec is a unit of distance. If megaparsecs are measured in kilometers, then the dimensions of the Hubble parameter would be kilometers per second per kilometer, or sec^{-1}. Sec^{-1} is a measure of frequency so that the Hubble parameter is a frequency measurement, which is actually the inverse of the age of the universe. We will refer to this Hubble frequency as ω_R, for reasons which will soon become apparent.

This accelerated expansion is a direct result of the decrease in the universal gravitational acceleration $\left(\frac{Gdm}{(dr_U)^2} \right)$ as mass (dm) moves across event horizons into black holes, and in fact, the rate of change in the universal gravitational acceleration with respect to the change in the Hubble parameter is exactly equal to the rate of change in a photon's electric field with respect to the change in its magnetic field; that is to say, it is equal to the speed of light (c).

$$\left(\frac{Gdm}{(dr_U)^2} \right)\left(\frac{1}{d(\omega_R)} \right) = c \qquad (11.53)$$

This is because

$$dm = \left(\frac{(F_p)d(r_U)}{c^2} \right) = \left(\frac{c^4}{G} \right)\frac{d(r_U)}{c^2}$$

where F_p is the Planck force. Therefore

$$\left(\frac{G}{d(r_U)^2} \right)\left(\frac{c^4}{G} \right)\left(\frac{1}{c^2} \right)\left(\frac{d(r_U)}{d(\omega_R)} \right) = \left(\frac{G}{G} \right)\left(\frac{c^2}{d(r_U)d(\omega)_R} \right) \qquad (11.54)$$

and

$$(d(r_U)d(\omega_R) = c$$

so that

$$\left(\frac{G}{d(r_U)^2}\right)\left(\frac{c^4}{G}\right)\left(\frac{1}{c^2}\right)\left(\frac{d(r_U)}{d(\omega_R)}\right) = \left(\frac{G}{G}\right)\left(\frac{c^2}{c}\right) = c \qquad (11.55)$$

As more mass falls into black holes, Fig. (26) changes accordingly, and each change is mirrored exactly as a set of waves from the frequency spectrum in Fig. (23). In fact Eq. (11.49) reflects these changes in the values of γ. ω_1 in Eq. (11.20) corresponds to a real γ (as apposed to an imaginary γ), so we will refer to it as ω_R, as apposed to ω_i; and in the equations above and below, "ω_R" *actually represents the Hubble parameter*. That is to say that the Hubble parameter is the inverse of a "point" in time, and corresponds precisely with the angular frequency of one of the waves in Fig. (23). Meanwhile, the universe expands at a faster rate, and time moves forward more quickly, as time goes on (in agreement with the fluctuation theorem, by the way).

And it follows that

$$\frac{t}{t_0} = \frac{r}{r_0} = \frac{m_0}{m} = \sqrt{1 - \left(\frac{\omega_R}{\omega_p}\right)^2} \qquad (11.56)$$

where ω_R is the Hubble parameter. Remember that, with respect to general relativity, the inertial frame of reference (the one with the zero subscript) is that frame which is located at an infinite distance from a given mass, so that in terms of the evolution of the universe, that frame would be represented by a real universe completely devoid of any mass. At that point, the Hubble parameter would be zero, and

$$\frac{t}{t_0} = \frac{r}{r_0} = \frac{m_0}{m} = 1$$

Comparing that universe with the one somewhat earlier in time, when the Hubble parameter was non zero, we see that, compared to the inertial frame, time (t) was earlier, the universal radius (r) was smaller, and the universal mass (m) was larger, so that

$$\frac{t}{t_0} = \frac{r}{r_0} = \frac{m_0}{m} < 1$$

And since

$$1 - \left(\frac{\omega_R}{\omega_p}\right)^2 = \left(\frac{\omega_p{}^2 - \omega_R{}^2}{\omega_p{}^2}\right) = \frac{(\omega^2)_{diff}}{(\omega_p)^2} \tag{11.57}$$

$$\frac{t}{t_0} = \frac{\omega_{dif}}{\omega_p} \tag{11.58}$$

where ω_p and ω_{dif} are like two clocks whose second hands are spinning at different frequencies. And likewise

$$\frac{r}{r_0} = \frac{\omega_{dif}}{\omega_p} \tag{11.59}$$

and

$$\frac{m_0}{m} = \frac{\omega_{dif}}{\omega_p} \tag{11.60}$$

where again, ω_{dif} is the square root of the difference in the squares of the frequencies.

With this in mind, Eq. (11.50) and Eq. (11.51) look like

$$\frac{t_1}{t_0} = \frac{\omega_{dif}}{\omega_p} = \sqrt{1 - \frac{\frac{dm_1}{dr_1}}{\frac{m_p}{r_p}}} \qquad \frac{t_p}{t_0} = \frac{\omega_{dif}}{\omega_p} = \sqrt{1 - \frac{\frac{m_p}{r_p}}{\frac{m_p}{r_p}}} \qquad \frac{t_2}{t_0} = \frac{\omega_{dif}}{\omega_p} = \sqrt{1 - \frac{\frac{dm_2}{dr_2}}{\frac{m_p}{r_p}}}$$

and

$$\frac{r_1}{r_0} = \frac{\omega_{dif}}{\omega_p} = \sqrt{1 - \frac{\frac{dm_1}{dr_1}}{\frac{m_p}{r_p}}} \qquad \frac{r_p}{r_0} = \frac{\omega_{dif}}{\omega_p} = \sqrt{1 - \frac{\frac{m_p}{r_p}}{\frac{m_p}{r_p}}} \qquad \frac{r_2}{r_0} = \frac{\omega_{dif}}{\omega_p} = \sqrt{1 - \frac{\frac{dm_2}{dr_2}}{\frac{m_p}{r_p}}}$$

Our present universe is mostly empty space, so that ω_R for our present Hubble parameter would be an extremely small number and $\frac{\omega_{dif}}{\omega_p}$ would be very close to 1 on average throughout the universe. As a result, the average curvature of space in the present universe is near zero, which is to say that space is very nearly flat and getting flatter all the time.

For a very large Hubble parameter, with ω_R approaching ω_p, which would signify a point near the beginning of time, ω_{dif} approaches zero so that

$$\frac{t}{t_0} = \frac{r}{r_0} = \frac{m_0}{m} = \frac{\omega_{dif}}{\omega_p} \to 0$$

In fact, it would appear that entropy would vary directly with ω_{dif}.

Meanwhile, at any instant in time corresponding to ω_R, where ω_R is the Hubble parameter and dm is the total mass which has fallen into black holes,

$$dm(r_p)^2\omega_R = I\omega_R = \hbar \tag{11.61}$$

where I is moment of inertia. Likewise, in the situation, where dm is that same mass and dr is the resulting change in the universal radius,

$$(dm)(c)(dr) = \left(\frac{dr}{r_p}\right)^2 \hbar \tag{11.62}$$

When all positive mass has moved across an event horizon Fig.(25) will look something like this:

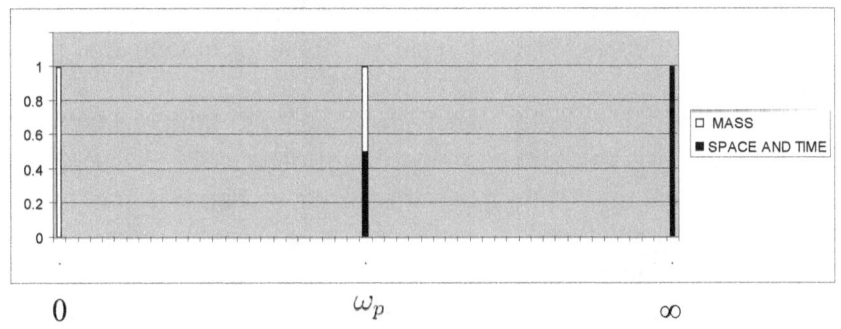

FIG. (37)

But these are just the first, center, and last vectors of Fig. (23). The superposition of the three is still just the Planck wave, the zero vector.

Why the Planck Wave

An obvious question at this point is, of course, "why the Planck wave". For instance, Eqs. (11.1)-(11.5) would seem to yield consistent results for any given set of waves, that is to say, for any arbitrary selection of ω_0. On the other hand, Eq. (11.10) yields a different constant for every possible choice of ω_0. When $\omega_0 = \omega_p$ that constant

$$k = \frac{m_p}{r_p} = \frac{c^2}{G}$$

is, in fact, just the gravitational constant, expressed in units of energy rather than units of mass; i.e., joules per cubic meters per second per second, rather than kilograms per cubic meters per second per second. Therefore the selection of ω_0 in Eq. (11.12) is explicitly the selection of the gravitational constant. It is not coincidental that the gravitational constant coincides with the zero vector. In fact, it is the gravitational constant that establishes the universal relationships among space, time, and mass.

In dimensional analysis, what are called the "dimensions of an equation", are always some combination of distance (i.e. space), time, mass, electric charge, or temperature. Electric charge is measured in coulombs. A coulomb is, in fact, just a number—a very large number—the number of elementary charges which, when moved through a potential difference of one volt, produces one joule of energy, which in turn is measured in dimensions of space, time, and mass. Temperature, on the other hand, is a measure of the average kinetic energy of a system of particles, and kinetic energy, again, is measured in dimensions of space, time, and mass.

It is probable, in fact, that the science of mathematics originated in an effort to keep track of the dimensions of space, time, and mass, most likely in commercial type enterprises on a fundamental level. Trading products of different sorts required keeping track of units of mass. Trading labor for products probably required keeping track of time. Tool making and building required the use of basic units of distance.

Understandably, most of the early spatial dimensions were based on the human body, many of which are still in use, in some form, today: the foot, the yard (approximately one human stride), the meter (likewise, one stride), the fathom (outstretched arms). The fluid ounce of the empirical system of measurements was, by the way, equal to two mouthfuls in the old English system. The time measurement, that we use today, is the approximate time between human heartbeats. Early units of mass were mostly based on the commodities being traded, and again, some are still in use today, the grain, for instance.

With the age of enlightenment, and especially with the advent of international trade, the need for standardization became more and more a necessity. Around 1790, a proposal was made by a Sir Christopher Wren, a founder of the royal society, that a system of measurement be based on the yard and a pendulum beating at one beat per second in the Tower of London. A similar proposal was later made by a Frenchman (Marie Jean Antoine Nicolas de Caritat Condorcet) that a system based on the meter and a pendulum second be adopted. In 1791, Jean Charles de Borda, a member of the French Academy of Sciences, proposed, insightfully, a system based on the *length* of a pendulum keeping time. In hindsight, this could have produced a system based on the earth's radius, the earth's mass, pendulum time, and the gravitational constant, all of which are more or less constant, and all of which are interrelated.

In the end, however, the metric system, which is in use in most of the world today, was passed into law by the French national assembly, which included a meter bar based on the distance between the north pole and the equator, the second, and a one-kilogram-metal weight. The gram, of course, is the mass equivalent of one cubic centimeter of water; therefore the meter, second, and kilogram were all based upon some aspect of the human anatomy. Over the years the metric system has been refined and fine-tuned. Today, the meter is defined as the distance light travels in a vacuum in $\frac{1}{299792458}$ of a second, and one second is defined as 9,192,631,770 cycles of radiation at a particular wavelength from a cesium-133 atom. The units of space and time therefore, can now each be defined in terms of the other, based upon a natural physical constant (the speed of light). The unit of mass however, is still defined in human terms, and cannot be related, in any way, to those of space and time.

The swing rate of a pendulum is directly proportional to the square root of the length of the pendulum and inversely proportional to the square root of the gravitational acceleration. Stop now and consider, instead, a system based on the solar year and a hypothetical or *imaginary* pendulum, consisting of a massive bob suspended from the earth on a massless tether so as to swing through the center of the sun. (The recently proposed "space elevator" would operate on a similar principle). The pendulum period t in this case, would be

$$t = 2\pi\sqrt{\frac{r}{\frac{(G)(m_s)}{r^2}}} = 2\pi\sqrt{\frac{r^3}{(G)(m_s)}} = 3.156 X 10^7 sec \qquad (12.1)$$

$$= 365.25 days$$

where r is the earth's orbital radius, m_s is the sun's mass, and G is the gravitational constant. Such a system would be based on the solar year, the sun's mass, and the earth's mean orbital radius, thus eliminating the anthropocentric aspect of the measurement system, and incorporating the gravitational constant. [5] Solving for G in Eq. (12.1)

$$G = \frac{r^3}{\left(\frac{t}{2\pi}\right)^2 (m_s)}$$

If we want to convert both sides of the equation to units of energy, as apposed to units of mass, we do so by dividing by the square of the speed of light, so that

$$\frac{G}{c^2} = \frac{r_e^3}{c^2\left(\frac{t}{2\pi}\right)^2 (m_s)} = \frac{1481m}{2.00 X 10^{30} kg} = \frac{r_p}{m_p}$$

---

[5]   In point of fact, such a measurement system would be impractical over time, since the sun's mass, the earth's orbital radius and the solar year are slowly changing.

Which of course, is the inverse of Eq. (11.13), where m_s corresponds to some dm_2 in Fig. (23), and $1481m$ corresponds to some dr_1. From this we could derive Eq. (11.22), and solve for (t) as in Eq. (11.26), in which case quantum gravity predicts the outcome of an exercise in Newtonian physics.

In fact, if

$$\gamma m_0 c^2 - m_0 c^2 = \frac{1}{2} m_0 v^2$$

as in Eq. (11.24), then

$$\gamma c^2 - c^2 = \frac{1}{2} \frac{m_0 v^2}{m_0} = \frac{v^2}{2}$$

and Eq. (11.25) becomes

$$\sqrt{\frac{4(\pi r)^2}{v^2}}$$

$$= \sqrt{\frac{4(\pi)^2 r}{\frac{v^2}{r}}}$$

$$= 2\pi \sqrt{\frac{r}{\frac{Gm_s}{r^2}}} \tag{12.2}$$

which is Eq. (12.1).

The point of all this is to show that the Planck wave (the zero vector), determines a universal system of measurement (at least for our particular universe), by establishing the relationships among space, time, and mass at the most fundamental level.

At this point, we should probably go back and try to consolidate all of this. Euclid said that "two things equal to the same thing are equal to each other". If this be the case, then "nothing" is equal to anything that is equal to "nothing". Euler pointed out that "nothing" was equal to $e^{i\pi} + 1$. From that we showed that

$$0 = (cos(Q\pi) + i\sin(Q\pi)) - (-1^Q)$$

$$= (\cos(Q\pi) + i\sin(Q\pi)) - (\cos(Q\pi) + i\sin(Q\pi))$$

an infinite set of complex unit vectors.

We saw that, from "nothing", we were able to generate the complete set of trigonometric, inverse trigonometric, hyperbolic, and inverse hyperbolic functions, the infinite set of integers, the infinite set of rational numbers, the infinite sets of real and imaginary numbers, an infinite set of irrational numbers, and at least a large set of transcendental numbers, thus demonstrating the tremendous complexity, and infallible logic of "nothing".

For that matter, if "nothing" contains the infinite set of dependent sets, then Fig (23) has to be in that set. Fig (23) represents an infinite number of waves, each of which is a unit vector. Out of that infinite set of unit vectors there is one and only one which is the zero vector, and that is the Planck wave. That, once again, is because

$$0 = m_p c^2 - \frac{G m_p^2}{r_p}$$

Meanwhile, the remaining vectors in the set (every one a non zero vector) form a linear combination that is equal to the zero vector, i.e., the Planck wave.

Fig (23) represents an infinite set of waves, part real and part imaginary. This infinite set of waves, of which "nothing" is comprised, spontaneously generates two sets of forces (Eqs. (11.43) and (11.44)), the respective members of which are initially equal and opposite (the zero vector). Each set consists of a constant force, and a force which varies directly with the product of two masses and inversely with the distance between the masses: a Planck force and a gravitational force.

These two forces permeate the universe, interacting with everything that possesses mass or energy. Curiously, it is the break in the symmetry between these two equal and opposite forces which seems to have initiated the universal expansion, beginning a 13.7 billion year odyssey in which the universe separates into real and imaginary parts, our universe expands, and time, as they say, "marches on".

Index

Note: Page locators with the letter *f* refer to figures.

www.ingramcontent.com/pod-product-compliance
Lightning Source LLC
Chambersburg PA
CBHW022059170526
45157CB00004B/1404